Practical
Fungal Physiology

Practical Fungal Physiology

Peter M. Robinson
Botany Department
The Queen's University of Belfast
Belfast

JOHN WILEY & SONS
Chichester · New York · Brisbane · Toronto

Library of Congress Cataloging in Publication Data:

Robinson, Peter M.
Practical fungal physiology.

Includes bibliographical references.
1. Fungi – Physiology. 2. Mycology – Experiments. I. Title. [DNLM: 1. Fungi – Physiology. QK601 R663p]
QK601. R58 589'.2'041 78-4243

ISBN 0 471 99656 4

Typeset in IBM Press Roman by
Preface Ltd, Salisbury, Wilts
Printed in Great Britain by
Unwin Brothers, The Gresham Press, Old Woking, Surrey

Contents

Preface

The potential of fungi as experimental material is not always fully exploited. Many fungi grow rapidly and lend themselves to experiments which can be carried out in minutes. In this respect they have advantages over many higher plants as tools for physiological research. The fungal species which feature in this book have been selected with a view to their ease of culture and handling. With one exception (the experiment on chemostat culture of fungi) experiments which require elaborate apparatus and facilities have been avoided.

This book is intended as a source of suggestions for experiments that might be included as part of an undergraduate course in mycology. It is also anticipated that some of the experiments may be suitable as a basis for ideas for projects at the sixth-form level of schools.

I would like to thank my colleagues in the Botany Department at Queen's University for their support during the preparation of this book. I also wish to express my thanks to the journals and publishers who have given permission to reproduce previously published material. Thanks are especially due to David Park for his encouragement and helpful criticism of the manuscript. Finally, many of the experiments concern problems which are as yet unanswered. It is hoped that in the course of carrying out some of the experiments other experiments will suggest themselves in addition to alternative hypotheses for the effects observed.

Belfast
1978

P.M.R.

1
Spore Germination

1.1 THE GERMINATION PROCESS

There is one general characteristic which distinguishes the spore phase from the vegetative phase of a fungus. This is the lower metabolic activity of the spore as compared with a mature hypha. Respiration occurs in resting spores and the onset of germination can often be detected by an increase in oxygen consumption. Materials for the fundamental metabolic processes of living cells are incorporated in spores during their formation but spores contain only a small amount of water, not enough to allow metabolism to proceed at more than a minimal rate.

Spores of most species of fungi germinate readily when provided with suitable nutrients. A source of carbon and nitrogen is usually required as well as a suitable temperature and pH for germination. The first visible sign of germination is an increase in size of the spore which becomes readily stainable with cytoplasmic stains, presumably due to increased permeability. Vacuoles appear in the cytoplasm of the spore. The emergence of a germ-tube can occur from anywhere on the spore surface in some species but from only one or more particular areas in other species. These areas are designated germ pores.

The early stages of spore germination have been studied for sporangiospores of *Rhizopus arrhizus* (Ekundayo and Carlile, 1964; Ekundayo, 1966). The increase in spore diameter with time is linear and the spore undergoes a twentyfold increase in volume prior to emergence of the germ-tube. Similar results have been obtained for many other species although the degree of swelling may vary considerably. A procedure for studying dye uptake and spore swelling in *R. arrhizus* is outlined in Flowsheet 1 and the results of a class experiment are given in Fig. 1. Approximately 90 per cent of the sporangiospore population is stained within one hour of exposure to a solution of methylene blue but germination does not commence until a further hour has elapsed. This experiment can be extended by measuring the diameter of each of a sample of sporangiospores at intervals of half an hour or so. The average diameter will be found to increase linearly with time. When carrying out the experiments note particularly any changes in shape which occur as the sporangiospores swell. Is there a correlation of cell shape with a change in permeability such that a sporangiospore only stains when it is a particular shape? What is peculiar about sporangiospores which do not swell and do not stain?

The examination of germinating spores of many species by the electron microscope has shown that a new wall or walls form inside the existing spore wall

Flowsheet 1. Uptake of methylene blue by sporangiospores of *Rhizopus arrhizus*

Sporangiospores of *R. arrhizus* are inoculated on a malt agar slope and incubated for 6 days at 25 °C.

↓

Pour 5 ml sterile distilled water on the slope, shake gently for 10 s and decant the spore suspension into a sterile McCartney bottle. Shake the suspension vigorously to break up clumps of spores.

↓

Spread loopfuls of spore suspension over Cellophane squares placed on S agar plates. Incubate at 25 °C and remove squares of the Cellophane at 0.5 h intervals, mount in methylene blue (1 per cent w/v in water) for 5 min and examine with the microscope for swelling, uptake of the stain and germination.

Notes:
1. Preparation of media. *Malt agar medium:* malt extract, 30 g; agar, 12 g (or as recommended by manufacturer); distilled water to 1 l. *S agar medium:* glucose, 0.7 g; $MgSO_4 \cdot 7H_2O$, 0.5 g; KH_2PO_4, 0.2 g; NH_4NO_3, 0.1 g; agar, 12 g; distilled water to 1 l. Sterilize for 15 min at 15 p.s.i. and pour approximately 10–15 ml into each sterile Petri dish (plastic or glass).
2. The Cellophane recommended is PT 300 (The British Cellophane Company, Regal House, Twickenham, Middlesex, England). Squares (approximately 2 cm square) are cut and then boiled in distilled water for 1 min. Transfer the squares in fresh distilled water to a crystallizing dish and autoclave for 15 min at 15 p.s.i. prior to use.
3. Determine spore diameter, if required, with a calibrated eyepiece micrometer (see Flowsheet 8). It is sufficient to measure a sample of 25 spores. Record germination and uptake of methylene blue for a sample of 100 spores.
4. Glass universal containers (28 ml capacity) can be used in place of McCartney bottles. Both types of container have aluminium screw-caps with rubber liners and can be autoclaved. The universal container has a wider neck than a McCartney bottle and may be considered more convenient to use, particularly in subsequent experiments which involve the use of pipettes for dilution operations.

during the period of swelling. This formation of new wall material has been described for *Rhizopus stolonifer* and *Rhizopus sexualis* (Hawker and Abbott, 1963), *Cunninghamella elegans* (Hawker *et al.*, 1970), *Cunninghamella echinulata* (Khan, 1975), and *Botrytis cinerea* (Gull and Trinci, 1971). In *B. cinerea* three new layers of wall material are formed prior to germination.

The process of spore germination can be studied most readily in those species of fungi which have large spores and whose spores germinate rapidly. Two such species are *Cunninghamella elegans* and *Geotrichum candidum*. The sporangiospores of the former species are spherical in shape and covered with long spines whereas the arthrospores of *G. candidum* are mainly cylindrical in shape with convex ends to the cylinders. Sporangiospores of *C. elegans* germinate by the production of one or more narrow germ-tubes which are sometimes difficult to observe when they first

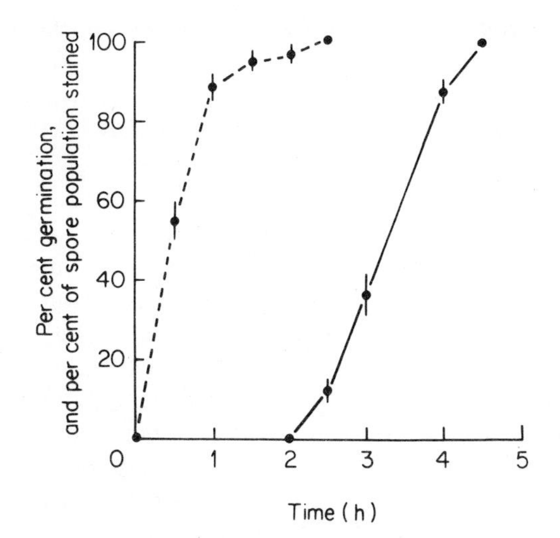

Fig. 1. Uptake of methylene blue by sporangiospores of *Rhizopus arrhizus* on S agar at 25 °C (– – –). The time-course of germination is also shown (——). Two hundred and fifty spores were recorded for each point on the graphs. Vertical lines represent 95 per cent confidence intervals for each value. (Data taken from a class experiment)

appear as they are surrounded by the spines on the outer spore wall. Arthrospores of *G. candidum* germinate by the production of a single germ-tube from one of the two convex ends of the cylindrical arthrospore. A second germ-tube can be formed at a later stage of growth.

Suitable procedures for following the time-course of germination in these two species are described in Flowsheet 2 and typical time-course curves for germination are shown in Fig. 2. The procedure outlined in Flowsheet 2 can be modified by seeding one plate with spores and examining the same plate at the recommended times. While satisfactory results can be obtained this way there is a tendency for the rate of germination to be reduced due to the length of time for which the plate will be at a lower temperature for recording purposes. An alternative procedure would be to seed one plate and remove a different small block of the seeded agar for examination at each of the intervals stated.

The time-course of germination for both species can also be determined in liquid media. In this case the spores from either species can be washed from a malt agar slope culture with 10 ml sterile malt broth into a small (100 ml capacity) sterile conical flask plugged with cotton wool. The suspension is shaken gently in a water bath at 30 °C and a sample (one drop) removed at intervals, placed on a glass slide and covered with a coverslip. Germination is recorded by using a microscope with a x 20 objective.

Flowsheet 2. Time-course of germination for spore populations of *Cunninghamella elegans* and *Geotrichum candidum*

A malt agar slope culture of each fungus is grown at 25 °C for 3 days. Seed the complete surface of the slope.

G. candidum

Pour 10 ml sterile distilled water on the slope, shake gently for 10 s and decant the spore suspension into a sterile McCartney bottle. Either streak a loopful of suspension over each of six S agar plates or pipette 0.2 ml suspension on each plate and spread evenly.

↓

Incubate plates at 30 °C. Examine one plate after 1.5 h and record the presence or absence of a germ-tube for each of 100 spores. Remove another plate after a further 0.25 h and assess germination again. Continue to remove a fresh plate at 0.25 h intervals and record germination. Draw a graph of your results.

C. elegans

Pour 4 ml sterile distilled water on the slope, shake vigorously for 30 s and decant the spore suspension into a sterile McCartney bottle. Shake the suspension vigorously for 1 min and either streak a loopful of suspension over each of six S agar plates or pipette 0.2 ml suspension on each plate and spread evenly.

↓

Incubate plates at 30 °C. Examine one plate after 0.75 h and record the presence or absence of a germ-tube for each of a sample of 100 spores. Remove another plate after a further 0.25 h and assess germination again. Continue to remove a fresh plate at 0.25 h intervals and record germination. Draw a graph of your results.

Notes:

1. Germination is best recorded by using a microscope with a x 20 objective. There will be a tendency for the objective to mist over but this can be avoided by placing a coverslip over the area of the plate to be studied.
2. Only record the presence or absence of germination in spores which are not in contact with other spores, i.e. avoid spores in clumps.
3. Take care not to record germination in a biased manner. It is useful to have an eyepiece graticule in the form of a grid such that every eligible spore in a given area is recorded and the same spore is not recorded more than once.
4. Reasonable results can be obtained by counting fewer than 100 spores in each sample but it is not recommended that fewer than 50 spores be recorded for each observation.
5. Preparation of media. *Malt broth:* formula as for malt agar (Flowsheet 1) but exclude agar. Other media as in Flowsheet 1.
6. Whenever a spore suspension is to be spread over the surface of a solid medium there is an advantage in incubating the plates at 35 °C for 24 h before they are inoculated. This promotes the rapid absorption of the liquid in which the spores are suspended and prevents movement of the spores during examination.

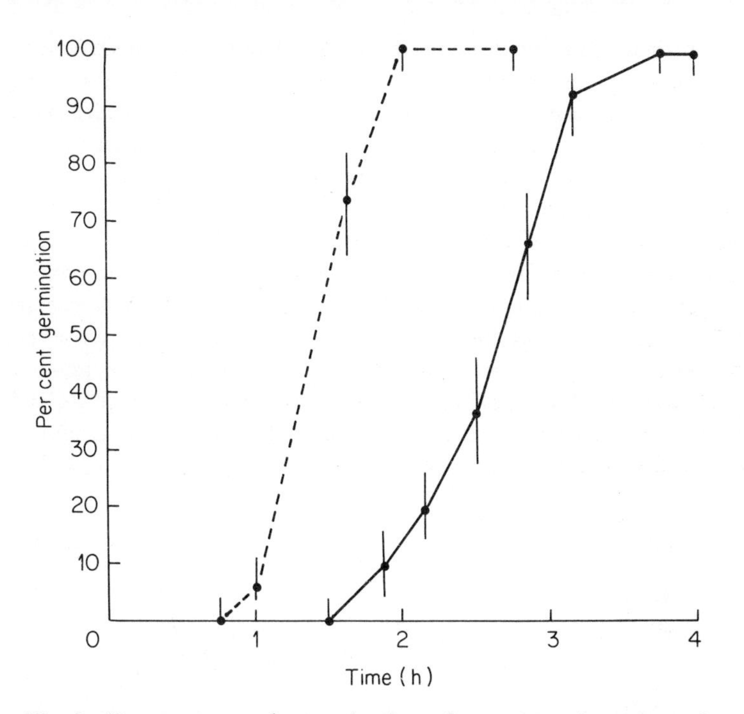

Fig. 2. Time-course of germination for sporangiospores of *Cunninghamella elegans* (– – –) and arthrospores of *Geotrichum candidum* (——) on S agar at 30 °C. One hundred spores were recorded for each point on the graphs. Vertical lines represent 95 per cent confidence intervals for each value. (Data taken from a class experiment)

1.2. EFFECT OF GROUP SIZE ON TIME-COURSE OF GERMINATION

Spores can interact in several ways such that a spore may behave differently when isolated from neighbouring spores. One difference is in the time taken for germination to occur and this effect can be conveniently studied using the procedure outlined for *Geotrichum candidum* in Flowsheet 2. When assessing germination, record germination for a sample of 100 arthrospores which are not in contact with any neighbouring arthrospores and also record germination for 100 arthrospores which are grouped in pairs such that each pair consists of two arthrospores in contact with each other but separated from other arthrospores. *G. candidum* is a very convenient organism for this experiment as the arthrospores are formed in chains and frequently remain attached to each other even when plated out. Express your results as a graph.

One problem which arises in this experiment is the time factor involved in recording such a large number of arthrospores at each selected time. Arthrospores can germinate during the recording period and this can exaggerate the germination percentage for the last sample to be recorded. There are several ways to overcome this bias. Students can work in pairs, one person recording the germination of single, isolated arthrospores, the other person recording arthrospore pairs. As an alternative procedure, the arthrospores and germlings could be killed at the appropriate times by placing a small volume (0.2 ml) of *n*-propanol in the lid of the inverted plate prior to recording.

Typical time-course of germination curves for single arthrospores and arthrospore pairs are shown in Fig. 3. In this experiment it is usual for the arthrospores in pairs to germinate more rapidly than the single arthrospores. This stimulation of germination occurs mainly in the first portion of the time-course curve for germination of the arthrospore population. There have been rare occasions when no difference was noted in the time-courses of germination for single and paired arthrospores but never an occasion when the single arthrospores germinated more rapidly than the paired arthrospores.

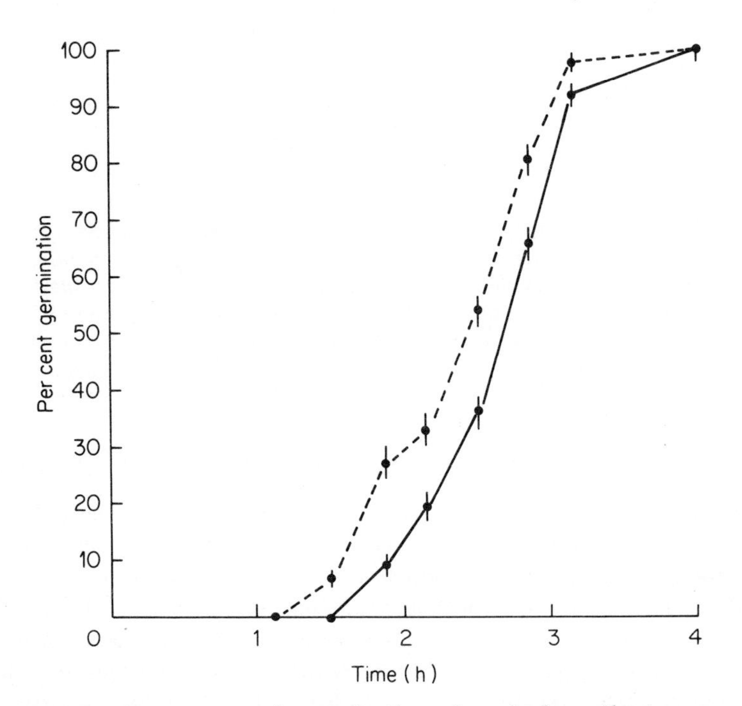

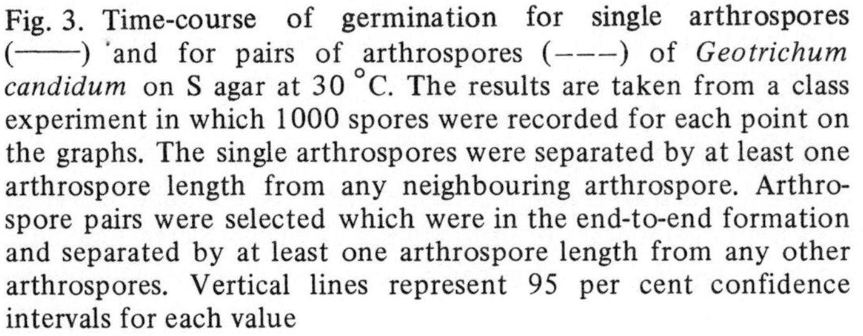
Fig. 3. Time-course of germination for single arthrospores (——) and for pairs of arthrospores (– – –) of *Geotrichum candidum* on S agar at 30 °C. The results are taken from a class experiment in which 1000 spores were recorded for each point on the graphs. The single arthrospores were separated by at least one arthrospore length from any neighbouring arthrospore. Arthrospore pairs were selected which were in the end-to-end formation and separated by at least one arthrospore length from any other arthrospores. Vertical lines represent 95 per cent confidence intervals for each value

One of the simplest hypotheses to explain these results is that each arthrospore produces a metabolite or metabolites which stimulate one or more of the various processes which culminate in germ-tube emergence. Such a factor would be more concentrated in a group of arthrospores. The experiment can be modified so that the time-courses of germination for groups of three, four, or five arthrospores are recorded. The larger groups point to a new problem. Whereas some of the arthrospores from the larger groups may germinate before any of the single arthrospores, a sample of 100 single arthrospores will usually germinate completely before a sample composed of large groups of arthrospores. In fact, some of the arthrospores in the larger groups may fail to germinate.

The effect of group size on the time-course of germination can also be conveniently studied in *Cunninghamella elegans*. A suitable procedure is outlined in Flowsheet 3. The first sporangiospores to germinate usually occur in the clumps.

Flowsheet 3. Effect of group size on time-course of germination for sporangiospores of *Cunninghamella elegans*

Inoculate a malt agar plate with spores from a 3-day-old slope of *C. elegans* grown on malt agar at 25 °C. Spread spores over the entire surface of the plate and incubate at 25 °C for 3 days.

↓

Place the inverted basal portion of the seeded plate over the basal portion of a plate of unseeded malt agar. Hold the two basal portions firmly together and bang them sharply on the bench. Several spores will fall from the seeded plate onto the exposed malt agar below and there will be several large clumps of spores. One plate of a 3-day-old culture of *C. elegans* can be used to inoculate several plates and experience will dictate how hard and how often to bang the paired plates to obtain a suitable sowing density.

↓

Incubate six seeded plates at 30 °C and at intervals of 0.75, 1, 1.25, 1.5, 2, and 3 h, take a fresh plate from the incubator and record germination of single (isolated) spores and spores in clumps of cells. Draw a graph of your results.

Notes:

1. If it is decided to study the group effect with spores of *Geotrichum candidum* use the procedure outlined at the beginning of this section. When the sample size is 100 spores significant differences may not be obtained between the germination percentages recorded for populations of single and paired spores. The stimulatory effect is usually reproducible when the sample size is increased to 500 or 1000 spores. When sample size is increased to this extent it is essential to seed several plates of medium and to use a separate plate for each observation. Immediately before observation invert the Petri dish and add 0.2 ml *n*-propanol to the inverted lid. This will prevent any further development of the spores and germlings which, assuming approximately 10 ml medium is used for each dish, can be examined through the base of the inverted dish.
2. Incubate all plates at 35 °C for 24 h prior to inoculation.

The germ-tubes of *C. elegans* grow rapidly and it is quite common to see very long germ-tubes growing away from the clumps at a time when many isolated sporangiospores have failed to germinate. This experiment raises an additional question. Is the effect of clumping mainly on stimulating earlier germination or is the rate of hyphal extension-growth also stimulated?

1.3 SELF-INHIBITION OF SPORE GERMINATION

Spores of many (if not all) species of fungi fail to germinate when seeded at high densities in media which would normally permit germination. In addition, several species of fungi have spores which are known to produce an inhibitor or inhibitors which prevent germination in an otherwise suitable medium. Allen (1955) demonstrated that such an inhibitor could be removed from urediospores of *Puccinia graminis* by floating them on an aqueous solution. This soaking procedure had to be performed under aerobic conditions and the urediospores germinated when resuspended in a fresh aqueous solution. Self-inhibition of germination in *P. graminis* attracted much interest but it was not until recently that the self-inhibitor was isolated and characterized as the *cis*-isomer of methyl ferulate (Macko *et al.*, 1971; Allen, 1972).

Lingappa and Lingappa (1965) studied the effects of nutrients on the self-inhibition of conidia of *Glomerella cingulata*. Washed conidia were distributed in distilled water in Petri dishes to give conidial densities ranging from 30 to 3000 mm^{-2} on the base of the dish (the conidia have a specific gravity greater than that of water). At a density of 30 mm^{-2} approximately 90 per cent of the conidial population germinated; at 3000 mm^{-2} less than 3 per cent of the conidia germinated. Germination of conidia was directly proportional to the density of the conidia in water. The decreased availability of oxygen did not seem to account for the inhibition of germination noted at the high conidial densities because shaking these suspensions did not result in more germlings than occurred in a static control dish. It was discovered that when conidia were washed five times or more with fresh changes of distilled water an improvement in germination occurred when compared with unwashed conidia seeded at a normally self-inhibitory density. Later work (Lingappa and Lingappa, 1966) showed that exudates from the conidia were inhibitory to germination. These exudates retained activity after autoclaving but the active principle has not been characterized.

Self-inhibition has been described for conidia of *Aspergillus nidulans* (Trinci and Whittaker, 1968). Conidia, after being washed in three changes of water, were resuspended at various densities in a synthetic liquid medium. At high densities (10^7 conidia ml^{-1} and higher) less than 50 per cent of the conidial population germinated. Germination of conidia was enhanced when the suspensions were aerated with air containing carbon dioxide or when sodium bicarbonate was incorporated into the medium. Germination was not enhanced, when compared with a non-aerated control, by aerating a self-inhibitory suspension with air from which the carbon dioxide had been removed. Self-inhibition observed in this species is probably due to an insufficient supply of carbon dioxide during the early stages of germination.

Self-inhibition of germination has also been studied in pycnidiospores of *Mycosphaerella ligulicola* (Blakeman, 1969) and in other organisms by other workers. It is not a phenomenon restricted to fungal spores since it can also be demonstrated in pollen grains and in seeds of higher plants.

Self-inhibition can be studied in *Geotrichum candidum* by sowing arthrospores at different densities as outlined in Flowsheet 4. The arthrospore densities have been calculated to give a range of germination in the population from 0 to 100 per cent in each of the three situations studied. A typical result is shown in Fig. 4.

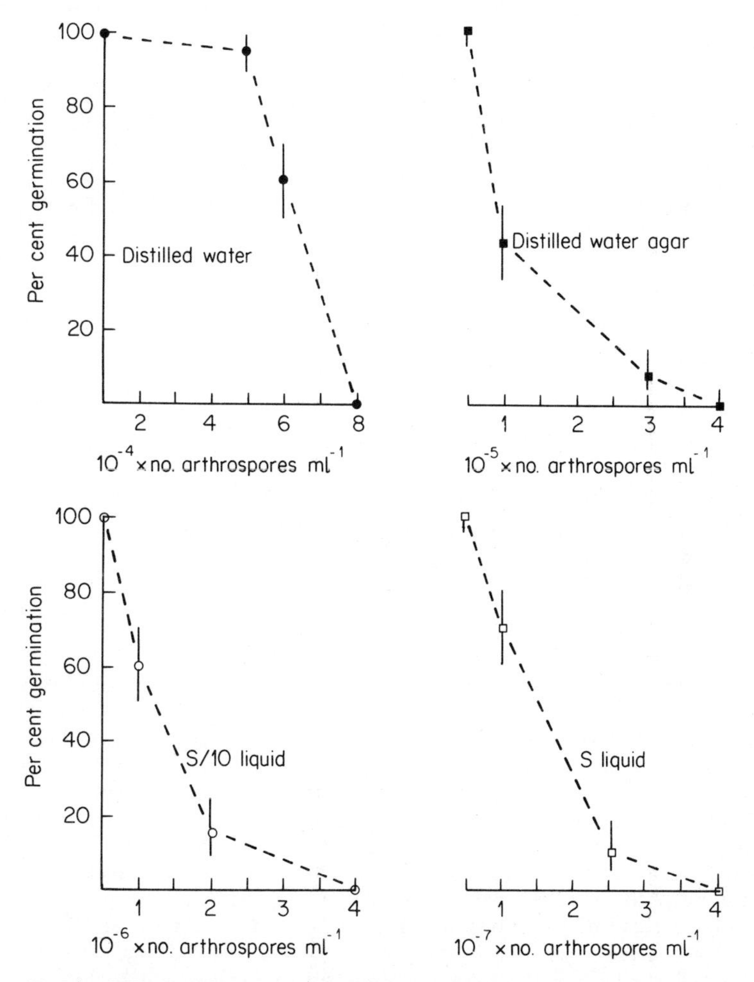

Fig. 4. Self-inhibition of germination of arthrospores of *Geotrichum candidum* in different media. Each point is the result of recording 100 arthrospores. Vertical lines represent 95 per cent confidence intervals for each value. (Data obtained from class experiments)

Flowsheet 4. Self-inhibition in arthrospores of *Geotrichum candidum*

Inoculate a medical flat slope of malt agar with *G. candidum* and incubate at 25 °C for 3 days. Add 5 ml sterile distilled water, shake gently for 10 s and decant the spore suspension into a sterile McCartney bottle. The spore suspension should have a density of approximately 5×10^7 spores ml^{-1} and this can be checked with a haemocytometer.

Spore suspension (5×10^7)

Left branch (1 ↙)	Centre branch (2 ↓)	Right branch (1 ↘)
○9 (5×10^6)	△8 (10^7)	○9 —1→ □9 (5×10^5)
1 ↓	1 ↓	2 ↓
○9 (5×10^5)	△9 (10^6)	○8 —1→ □9 (10^5)
2 ↓	1 ↓	5 ↓
○8 (10^5)	△9 (10^5)	○5 —1→ □9 (5×10^4)
5 ↓		
○5 (5×10^4)		
2 ↓		
○8 (10^4)		

Legend:

○ = no. ml sterile distilled water in McCartney bottle

□ = no. ml sterile distilled water agar in McCartney bottle

△ = no. ml sterile S/10 liquid in McCartney bottle

Numbers in brackets relate to spore densities (no. spores ml^{-1}); numbers *not* in brackets refer to volumes (ml) of spore suspension to be transferred.

Notes:

1. The medical flat slope of malt agar consists of a 200 ml capacity medical flat bottle to which is added sufficient medium (normally 30 ml) to cover the basal flat portion of the bottle when laid horizontally. After sterilization the bottle is laid on its side to allow the medium to set. It is suggested that the malt agar is prepared with 20 g agar l^{-1} to give a firmer consistency. The medical flat slopes can be inoculated by decanting a spore suspension from a McCartney bottle slope of *G. candidum* into the medical flat, allowing the suspension to run over the whole surface of the agar and decanting the suspension into a second medical flat and so on. Several medical flat slopes can be inoculated in this way from a single McCartney slope, sufficient spores remaining in each medical flat slope to give a uniform surface growth in the bottles which should be incubated horizontally.
2. For each spore suspension in distilled water and in S/10 liquid, 5 ml should be transferred to a small (5 cm diameter) sterile glass or plastic Petri dish. For each spore suspension in distilled water agar the agar is melted and held at 50 °C in a

Flowsheet 4. (continued)

---

water bath. The spore suspension is added, shaken gently, and poured into a 9 cm diameter Petri dish which is rocked gently on the bench to ensure the suspension covers the whole of the base of the dish.
3. Incubate all the dishes at 30 °C and after 4 h record the presence or absence of germination for a sample of 100 spores from each treatment. It is useful to record the dishes with suspensions in distilled water and in distilled water agar again at 24 h since sometimes germination is not complete in water at 4 h.
4. It will be noted that some spores float on the surface of the suspensions in distilled water and in S/10 liquid. Do not include any of these spores in the sample as they will frequently germinate whereas spores which have settled may fail to do so.
5. Preparation of media. *S/10 liquid:* formula as for S agar (Flowsheet 1) but exclude agar and dilute 10 times. *Distilled water agar:* agar, 12 g; distilled water to 1 l.

It will become apparent from the results of these experiments that arthrospores of *G. candidum* are able to germinate in distilled water and in distilled water agar and that the presence of nutrients increases the sowing density at which self-inhibition occurs. Strictly speaking, there will be a small carry-over of nutrients from the malt agar slope on which the arthrospores were seeded and this could invalidate the conclusion that the arthrospores are able to germinate in distilled water. However, if the decanted arthrospore suspension is washed with distilled water on a membrane filter prior to use very little difference is noted in the results obtained.

When self-inhibition has been demonstrated in one or more of the situations outlined in Flowsheet 4 attempts can be made to determine the mechanism of the effect. Some suggestions for experiments are described in Flowsheet 5 and the results of these can be used to construct simple hypotheses to account for the phenomenon. Arthrospores seeded in distilled water agar at a density of 5×10^6 arthrospores ml^{-1} do not normally germinate whereas arthrospores seeded at 5×10^4 arthrospores ml^{-1} will usually germinate completely. Experiment (1) tests whether or not arthrospores seeded in distilled water agar at the lower density are affected by contact with arthrospores seeded at the higher density. Experiment (2) tests whether a volatile factor is involved. Experiments (3) and (4) are designed to permit the detection of any factors diffusing through one or more layers of Cellophane or agar. It should be stressed that not only should the possibility that an inhibitor of germination is involved be considered but also that some factor essential for germination may be competed for by the high-density arthrospore population which may function as a sink for such a factor or factors.

It is by no means anticipated that these simple experiments will determine the mechanism of self-inhibition in *G. candidum*. The experiments have been selected primarily because they give reproducible results. It is hoped that some new ideas will be generated from these basic experiments and that these will serve as a foundation from which to launch a further attack on this complicated phenomenon.

Flowsheet 5. Mechanism of self-inhibition in arthrospores of *Geotrichum candidum*

Prepare a spore suspension with a density of 5×10^7 spores ml^{-1} using two medical flat slopes of malt agar and the technique outlined in Flowsheet 4.

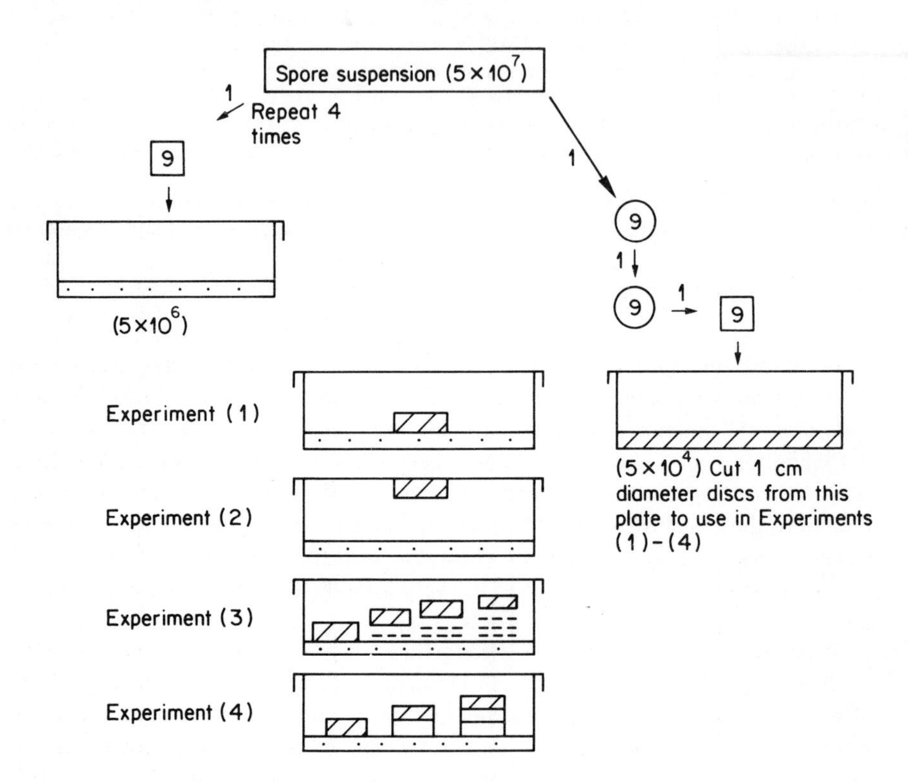

Legend:

○ = no. ml sterile distilled water in McCartney bottle

□ = no. ml sterile distilled water agar in McCartney bottle

Numbers in brackets relate to spore densities (no. spores ml^{-1}); numbers *not* in brackets refer to volumes (ml) of spore suspension to be transferred.

- - - = layer of Cellophane

[dotted box] = spores seeded in distilled water agar at a density of 5×10^6 spores ml^{-1}

[hatched box] = spores seeded in distilled water agar at a density of 5×10^4 spores ml^{-1}

[open box] = distilled water agar

Flowsheet 5. (continued)

Notes:

1. It is recommended that 20 g agar l^{-1} be used in all solid media as this facilitates cutting and transfer operations.
2. Cut discs from plates with a 1 cm diameter cork borer which is dipped in alcohol and flamed prior to use.
3. Prepare Cellophane as in Flowsheet 1.
4. Record germination for a sample of 100 spores from each treatment after 24 h incubation at 25 °C.
5. Whenever germination occurs in Experiments (3) and (4) measure the length of each of a sample of the germlings formed.

1.4 AUTOTROPISM IN GERMINATING SPORES

A tropism is a growth movement exhibited by part of an organism in response to a unidirectional stimulus. When the stimulus is due to the same organism which displays the tropism, or to a neighbouring organism of the same species, the oriented growth which occurs is an autotropism. Here we are concerned with autotropic responses which arise from interactions of neighbouring spores and which affect the point of emergence and initial direction of growth of each germ-tube. In the simplest sense there is an interaction of two spores but these responses may be modified by the density of the population to which the spores belong and also by other environmental conditions.

Jaffe (1966) studied autotropic effects of conidia from *Botrytis cinerea*. Dry conidia were suspended in portions of an inert fluorochemical liquid and electrostatically precipitated in polystyrene containers; the spores were then covered with a liquid medium to permit germination. The positions of germ-tubes of conidia in pairs were noted. It was found (Robinson *et al.*, 1968) that the autotropic behaviour of spore pairs of fungi could be conveniently studied by sowing spore suspensions on the surface of solid media. This technique minimized the risk of spores moving relative to each other as could happen on occasions when spores were studied in liquid media.

Three main types of autotropic behaviour may occur between neighbouring spores, namely, positive, negative, or neutral. Jaffe (1966) demonstrated that germinating conidia of *B. cinerea* showed pronounced positive autotropism (i.e. the spores germinate towards each other). The system used by Jaffe to classify the autotropic behaviour of spore pairs is illustrated in Fig. 5. On the other hand, it was found (Robinson *et al.*, 1968) that conidia of *B. cinerea* were neutral in autotropic behaviour when spore pairs seeded on the surface of solid media were examined. Sporangiospores of *Mucor plumbeus* were found to exhibit marked negative autotropism; sporangiospores of *Rhizopus stolonifer* and conidia of *Trichoderma viride* also showed negative autotropism but this was not so pronounced as in *M. plumbeus*. The patterns of germ-tube emergence exhibited by spore pairs of the different species of fungi examined indicate that a range of autotropic behaviour is likely to be found in fungi. Variation in autotropic behaviour of one species can be

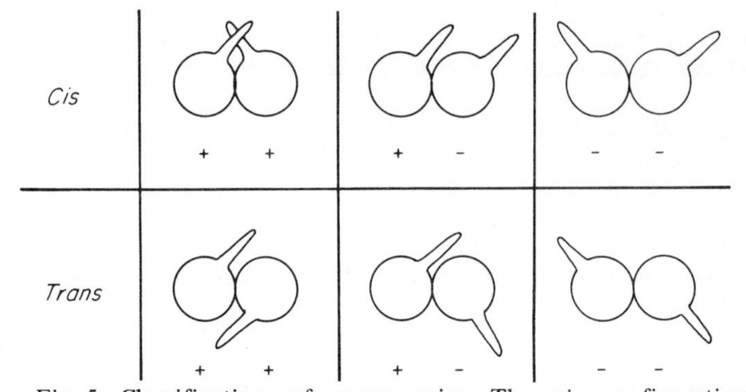

Fig. 5. Classification of spore pairs. The *cis* configuration occurs when both germ-tubes emerge on the same side of a plane perpendicular to the surface of the agar and passing through the centres of the two spores; when the germ-tubes emerge on opposite sides of this plane the spore pair is classified as *trans.* Spore pairs are also classified (+ +), (+ –), or (– –) with regard to the direction of growth of each emergent germ-tube; a germ-tube orientated towards the neighbouring spore is classified (+), and (–) when orientated away from it. In all species examined there were a few spore pairs which could not be classified according to the *cis, trans* (+ +, + –, – –) system. When the germ-tubes emerged directly away from or directly towards one another the spore pair could not be designated *cis* or *trans.* Similarly, when the germ-tubes of a spore pair emerged perpendicular to a line joining the centres of the spores they could not be recorded as (+) or (–). Such pairs were neglected because of their rare occurrence (After Jaffe, 1966). Reproduced by permission of the American Society of Plant Physiologists

marked under different experimental conditions but is very slight under standard conditions.

Although setting up experiments to observe autotropic behaviour can be relatively simple the measurement of the response can be difficult and require much patience. Many species of fungi produce spores which are spherical and in many instances the germ-tube(s) can be produced from any point on the spore surface. To record the point of germ-tube emergence precisely requires a protractor graticule for the microscope eyepiece so that the angle of emergence can be calculated with reference to an imaginary line linking the centres of the two neighbouring spores. The procedure in Flowsheet 6 enables autotropism to be studied for pairs of arthrospores of *Geotrichum candidum.* This organism has several attributes which make it an ideal subject for autotropic studies. Perhaps the main advantage is that when seeded on S agar each arthrospore germinates from one of its two ends. This simplifies observation and recording as there is no necessity to estimate the effect in terms of an angle – either an arthrospore germinates from the end next to its neighbour or from the opposite end. The experiment outlined in Flowsheet 6 should demonstrate the marked negative autotropism exhibited by arthrospores of this species and some typical germination patterns are shown in Fig. 6. Hypotheses

Flowsheet 6. Autotropism in arthrospores of *Geotrichum candidum*

Three-day-old malt agar slope of *G. candidum* grown at 25 °C.

↓

Pour 10 ml sterile distilled water on slope, shake gently for 10 s and decant spore suspension into a sterile McCartney bottle.

↓

Pipette 0.2 ml spore suspension on each of several S agar plates and spread the suspension evenly. Incubate at 30 °C for 2.5 h.

↓

Record the autotropic behaviour of 50 pairs of spores which are separated by at least one spore's length from any neighbouring spores.

Notes:

1. As a short-cut technique 10 ml sterile distilled water can be added to the slope of *G. candidum*, shaken, and the whole suspension poured on a plate of S medium. The suspension can then be poured from this plate onto a second sterile plate and so on. Surplus suspension can be drained from each plate and the plates incubated with the lids ajar to aid drying. With this technique the spore density declines as more plates are inoculated and the autotropic effects can be studied at these different densities.
2. The most usual way in which spores occur in pairs is with the short ends in contact. Sometimes pairs will be found with the long sides in contact (side-to-side) or in a T-formation (end-to-side). Autotropic behaviour can be studied for each of these patterns.
3. Spores will sometimes be found in chains (end-to-end) of more than two spores. Record the autotropic behaviour of such chains.

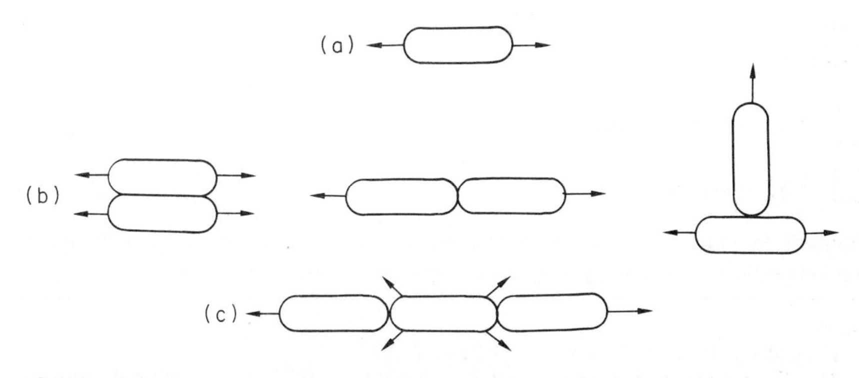

Fig. 6. Germination patterns in *Geotrichum candidum* on S agar. Each arthrospore normally produces a germ-tube from one of the positions indicated by the arrows. (a) Single arthrospore, (b) arthrospore pairs, (c) chain of arthrospores. Note the absence of germ-tube formation from the end of an arthrospore in contact with another arthrospore in the spore pairs (Robinson, 1973b). Reproduced by permission of Blackwell Scientific Publications Ltd

Flowsheet 7. Chemotropic response of arthrospores of *Geotrichum candidum* to oxygen

Prepare spore suspension of 5×10^7 spores ml^{-1} as described in Flowsheet 4 but wash spores from medical flat slope with sterile S liquid in place of sterile distilled water.

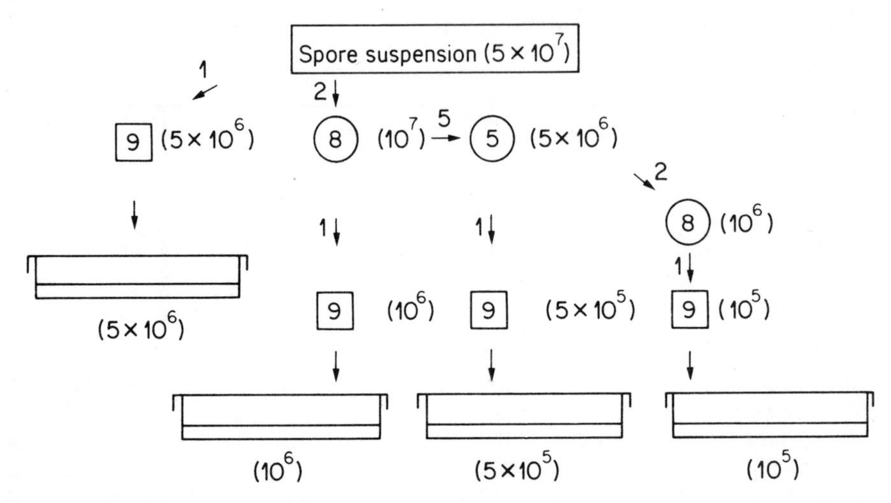

Experiment (1) Place a sterile glass coverslip on each plate.

Experiment (2) Place a plastic coverslip with a 1 mm diameter central perforation on a separate area on each plate.

Experiment (3) Place a second perforated plastic coverslip (separate from the first) on each of the plates seeded at 5×10^6 and 5×10^5 spores ml^{-1}. Remove a 1 cm diameter disc from each of these plates and place centrally on one of the two perforated plastic coverslips on the other plate. Place a glass coverslip over each disc.

Legend:

○ = no. ml sterile S liquid

□ = no. ml sterile S agar

Numbers in brackets relate to spore densities (no. spores ml^{-1}); numbers not in brackets refer to volumes (ml) of spore suspension to be transferred.

Notes:

1. Record results after 3 to 4 h incubation at 30 °C.
2. The S agar media should be melted and held at 50 °C in a water bath and the spore suspension added and mixed just before pouring into a sterile Petri dish. It is recommended that 20 g agar l^{-1} is incorporated into the S agar medium.
3. 1 per cent malt agar can be substituted for S agar in each of these experiments. This results in very clear-cut responses which occur at lower sowing densities than would be required to achieve a similar response in S agar. The explanation probably

Flowsheet 7. (continued)

lies in the fact that 1 per cent malt agar is a richer growth medium than S agar and would result in a higher rate of oxygen uptake by spores. As a result, for a given sowing density, spores seeded in 1 per cent malt agar and covered with a coverslip would reduce the oxygen tension in their immediate environment more rapidly than if seeded in S agar.

When spores of *G. candidum* are seeded in 1 per cent malt agar it is quite common, particularly at low sowing densities, for each spore to produce two germ-tubes. Also, the spores swell to become oval in shape prior to germination. When aeration is restricted by application of a coverslip the tendency to multiple germ-tube formation is reduced and, at high sowing densities, it is usual for a single germ-tube to be initiated from the region of the spore surface nearest to the source of air.

4. *S liquid medium:* formula as for S agar (Flowsheet 1) but exclude agar. *1 per cent malt agar:* malt extract, 10 g; agar 20 g; distilled water to 1 l.

can be formulated to account for the results obtained, the simplest being that each arthrospore produces a diffusible metabolite which retards one or more of the processes which result in germ-tube emergence. Such a factor would be reinforced in the region where neighbouring arthrospores are in contact. The experiments can be extended in many ways, one way being to study how the autotropic influence of neighbouring arthrospores is affected by the distance which separates them. The germination patterns of large irregular clumps of arthrospores can also be studied.

1.5 CHEMOTROPISM TO OXYGEN

The fact that germinating spores can respond to a gradient in oxygen concentration can be seen as having a considerable survival value for a species growing under conditions where a low oxygen tension prevails. This tropism is not normally apparent in germinating spores except when environmental conditions are such that the oxygen tension begins to limit germination and the subsequent growth of the germ-tube. Under these conditions the germ-tube may emerge on the side of a spore facing the higher concentration of oxygen and the subsequent growth of the germ-tube will be towards the source of the oxygen (Robinson, 1973a).

This tropism can be demonstrated in many fungi but with a view to a simple analysis of results arthrospores of *Geotrichum candidum* are convenient to study for the reasons already outlined in Section 1.4. Some experiments are suggested in Flowsheet 7 and some of the results expected are shown in Fig. 7. Several points emerge from the results of this exercise. Experiment (1) demonstrates that the arthrospores will germinate at all the densities to be studied in S agar. Under the glass coverslip, however, oxygen availability will limit germination of some of the arthrospores seeded at the higher concentrations. When oxygen concentration begins to limit germination a gradient in germination is apparent with germination restricted to those arthrospores which are situated at the periphery of the coverslip. The higher the sowing density the narrower the width of the band of arthrospores

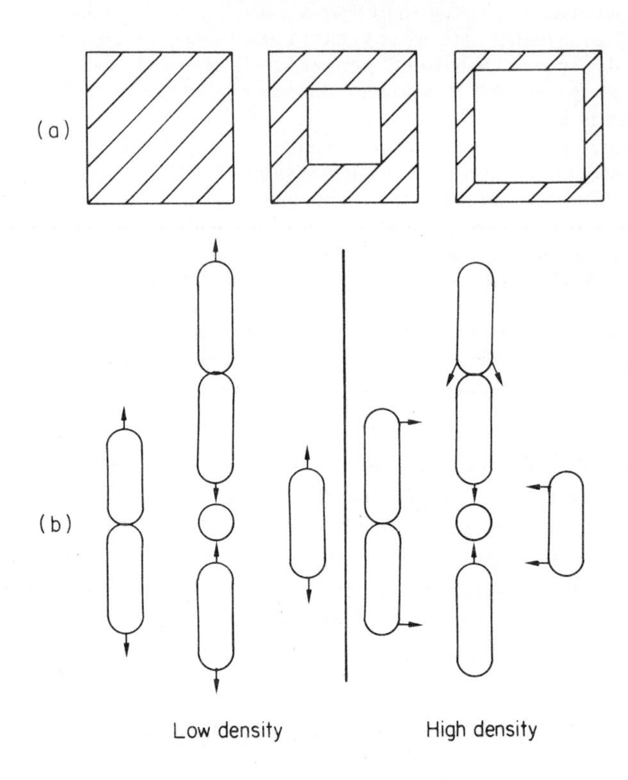

Fig. 7. Chemotropic responses to oxygen by germinating arthrospores of *Geotrichum candidum* seeded at different densities.

(a) Glass coverslip placed on surface of S agar seeded with arthrospores. Germination is restricted to the shaded area beneath the coverslip.

(b) Perforated plastic coverslip placed on surface of S agar seeded with arthrospores. Germination patterns of single and paired arthrospores are shown with respect to the central performation.

(c) High- and low-density seeded S agar discs separated by a perforated plastic coverslip.

→ = Sites of germ-tube formation and direction of germ-tube growth (Robinson, 1973b). Reproduced by permission of Blackwell Scientific Publications Ltd

which germinates and the greater the tendency for each arthrospore to germinate from the end nearer the margin of the coverslip. The lower the sowing density the greater the germination of the arthrospore population beneath the coverslip. Fewer arthrospores germinate from their ends nearer the margin but after some growth has been made marked tropic responses of the germ-tubes to the margin may occur. This is possibly due to the oxygen concentration becoming limiting for continued growth.

Experiment (2) enables arthrospore germination to be studied in the vicinity of a small hole in contact with the atmosphere. Experiment (3) demonstrates how spore populations of differing density compete for available oxygen and how the tropic responses of the two populations may differ in the vicinity of the perforation separating them.

2
Hyphal Growth

Before carrying out experiments on hyphal growth it is useful to consider some of the more general aspects of growth in fungi. The filamentous fungi grow by the production of filaments or hyphae which extend throughout the particular medium being colonized. When a fungus colony is grown on a solid medium the hyphae grow in a radial fashion from the inoculum and the resultant colony usually has a circular outline. The individual hyphae are cylindrical in shape and each tapers towards the distal (tip) region which is characteristically rounded in outline. Hyphal growth is referred to as 'extension-growth' and occurs at this tapered region which is expanding continuously due to the incorporation of new cell material.

Plomley (1959) studied the growth of hyphae of *Chaetomium* grown on a solid medium. He concluded that the organism had a functional growth unit which was defined as 'a free-growing hyphal tip associated with a growing mass of constant size in constant environment'. Plomley suggested that the growth unit of *Chaetomium* was about 150 μm long because each germ-tube produced a branch when it attained this length. The growth unit is defined as follows:

$$\text{Hyphal growth unit } (\mu\text{m}) = \frac{\text{Total hyphal length } (\mu\text{m})}{\text{Number of hyphal tips}}.$$

Trinci (1974) has shown that the number of hyphal tips possessed by a fungus colony increases exponentially with time at approximately the same specific growth rate as the total hyphal length of the mycelium. These results support the hypothesis that growth of a fungus mycelium or colony involves the duplication of a 'growth unit' which consists of a growing hyphal tip and a certain mean length of hypha.

The first-formed hyphae of a young colony gradually increase in extension-growth rate until the maximum extension-growth rate is established for each hypha under the particular culture conditions studied. The rate then remains constant unless growth becomes reduced by one or more environmental factors. Any lateral branches formed also reach a maximum rate of extension-growth but this is usually lower than that of the main hyphae. Henderson Smith (1924) measured the rate of extension-growth of individual hyphae from germinating spores of *Botrytis cinerea.* The plot of the log of the length of a main or lateral hypha showed an asymptotic curve but the total lengths of hyphae (main and lateral) gave a linear plot so that

Log Σ Total lengths of hyphae $\propto$ Time (Fig. 8).

Henderson Smith suggested that apical growth was regulated by the rates of uptake of nutrients and of their translocation to the hyphal apices.

It may be helpful at this stage to describe the 'peripheral growth zone' and 'hyphal differentiation'. As a colony spreads over a solid medium, growth is confined mainly to the apical regions of the leader (main) hyphae and their subapical and lateral branches. The length of a hypha which would be necessary to maintain its existing rate of extension-growth, or the width (in a radial direction) of the marginal region of a colony which can sustain the existing radial growth rate of the colony, is the peripheral growth zone. This is the subject of a later experiment (Flowsheet 12). When a colony spreads over a solid medium the process of hyphal differentiation occurs. In the initial stages of colony growth all the hyphae (main hyphae and lateral branches) are of a similar width and grow at a similar extension-growth rate. As the colony develops, a process of differentiation occurs whereby the leader hyphae have a faster rate of extension-growth than the lateral branches which in turn grow faster than any branches which they themselves bear.

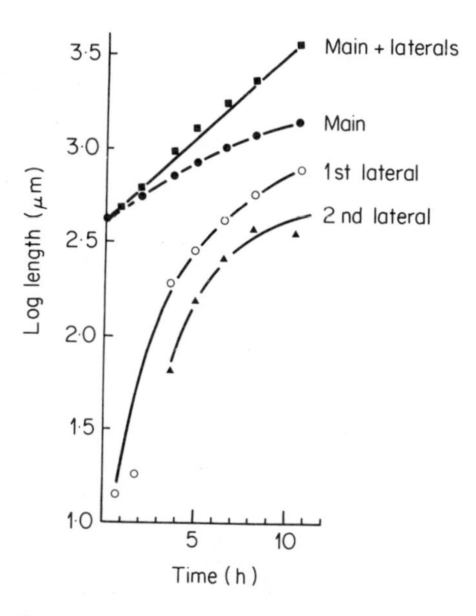

Fig. 8. Growth of hyphae of *Botrytis cinerea*. The extension-growth rate of a main hypha and lateral branch hyphae of germinating spores of *B. cinerea* were recorded for 10 h and log length plotted against time (Data from Henderson Smith, 1924). Reproduced by permission of Blackwell Scientific Publications Ltd

The faster-growing hyphae are also wider and usually have a longer tapered region at the hyphal tip.

An account of hyphal growth would be incomplete without reference to the actual mechanism by which extension-growth occurs. An understanding of the fine structure of the fungal hypha is also useful. Hyphal tips of fungi are of particular interest cytologically as growth is restricted to the rounded portion of the apex and cell components involved directly in cell expansion are likely to be concentrated in this region. Recent electron microscope studies have revealed that in all fungi examined the hyphal apex is filled with cytoplasmic vesicles, to the exclusion of almost all other organelles. These apical vesicles are interpreted as being secretory vesicles involved in wall elaboration and increase in plasma membrane at the growing apex. Such concentrations of vesicles are of general occurrence at sites of hyphal growth. In addition, septate fungi possess a specialized region within this apical zone. This region, which is distinct from the cluster of apical vesicles, corresponds to a small, refractile spheroid body – the spitzenkörper (tip-body) – observed by light microscopy in living hyphal tips.

The subapical hyphal zone contains all the normal protoplasmic components including mitochondria, nuclei, ribosomes, and an endomembrane system consisting of rough-surfaced endoplasmic reticulum, smooth-surfaced cisternae, and cytoplasmic vesicles. In Oomycetes, dictyosomes of the Golgi apparatus are the endomembrane components which produce vesicles. The non-oomycetous fungi do not have dictyosomes consisting of stacks of cisternae. Instead, each has in the subapical zone a characteristic endomembrane system which includes some form of smooth-surfaced cisternae or tubules which are associated with production of vesicles.

If we look with light microscopy (a good microscope is essential) at living hyphae of an Oomycete such as *Pythium* we can see an apical zone of rapidly-moving particles that correspond to the vesicles seen in electron micrographs. Grove and Bracker (1970) have described the situation in *Gilbertella persicaria* which has wide, rapidly-growing hyphae in which the individual vesicles are easily resolved. The vesicles exhibit a swarming motion. Those in the subapical region move along the cell wall towards the apex and appear to adhere to the inner surface of the wall at the hyphal apex. The wall at the apex of most hyphae has a more uneven profile than the lateral wall when viewed with the electron microscope and the undulations along the inner margin are about the same size as the apical vesicles. Profiles showing continuity between vesicle membranes and plasma membrane are seen in hyphal apices. With light microscopy, the apical vesicles are seen as a crescent-shaped band lining the apical wall. Individual vesicles at the apex disappear from view as the hyphae elongate and are constantly replaced by other vesicles from the subapical zone. The correlation of these apical vesicles and extension-growth is further emphasized by the fact that when growth ceases the zone of vesicles disperses soon afterwards.

The spitzenkörper found in septate fungi represents a specialized region within the cluster of apical vesicles. It is positioned near the central axis of the hypha and usually measures less than half the diameter of the hypha. The region lacks a distinct boundary and consists of either a vesicle-free area or an aggregate of very small vesicles or tubules. The spitzenkörper is seen as a roughly spherical, phase-contrast dark region and disappears as soon as growth ceases. When growth resumes the position of the spitzenkörper is correlated with the subsequent direction of hyphal elongation.

2.1 TIME-COURSE OF GROWTH

Fungal hyphae grow at rates which differ markedly from species to species. The rates of extension-growth of a range of fungal species are shown in Table 1. The Phycomycetes include some of the faster-growing fungi and species of *Saprolegnia* are particularly useful for studies on the time-course of growth. In contrast, *Aspergillus niger* or *Geotrichum candidum* can be examined to illustrate somewhat slower growth. The extension-growth of hyphae is best studied with the aid of time-lapse photography, the fungi being grown in enclosed, temperature-controlled chambers. A less precise but satisfactory technique is outlined in Flowsheet 8.

When measuring the rate of extension-growth of a hypha the hyphal-tip region may form a branch during the period of measurement. This can occur in *A. niger* and *G. candidum* but *Saprolegnia* spp. do not normally produce branches from the

Table 1. Rates of extension-growth of various fungi (Burnett, 1968). Reproduced by permission of Edward Arnold (Publishers) Ltd

	Growth rate	
	mm h^{-1}	μm min^{-1}
PHYCOMYCETES		
Mucor sp.	0.6–0.8	10–13
Phycomyces sporangia	3.9	65
Basidiobolus ranarum	0.26	4.4
ASCOMYCETES		
Gelasinospora tetrasperma	4.5	75
Humaria sp.	1.5–2.7	25–45
Neurospora crassa	1.5–6.0	25–100
Peziza sp.	1.1–2.0	18–33
FUNGI IMPERFECTI		
Penicillium sp.	0.01–0.02	0.17–0.33
Sclerotinia sp.	0.3	5
Botrytis cinerea	0.06–0.09	1–1.5
BASIDIOMYCETES		
Coprinus lagopus	0.10–0.25	1.7–4.2
Trametes vittata	0.28	4.7
Psathyrella disseminata	0.28	4.7
Schizophyllum commune	0.13–0.22	2.2–3.7

Flowsheet 8. Extension-growth of hyphae

Inoculate separate plates of 0.1 per cent malt agar with *Aspergillus niger, Geotrichum candidum*, and a species of *Saprolegnia*. Incubate plates for 2 days at 25 °C. Inoculate *Saprolegnia* at the edge of the plate and inoculate *A. niger* and *G. candidum* as diametric streaks.

↓

Align a reference point on an eyepiece micrometer with the tip of one of the marginal hyphae growing on the surface of the medium. As these measurements are usually made by direct observation of the lidless culture slight hyphal movements may occur due to drying out of the medium. It is useful to have a reference point on the hypha to be meaured such that a division on the eyepiece micrometer can be aligned with this point when measurements commence. The alignment should be checked from time to time. A suitable reference point would be the point on a hypha from which a lateral branch has emerged (see Fig. 10). Measure extension-growth (in eyepiece micrometer units) at intervals of 2 min for a period of 10 min or so. Calibrate the eyepiece micrometer using a stage micrometer and use the x 20 objective for all measurements if possible. Draw a graph of extension-growth against time for each species studied.

↓

Measure the extension-growth rate of lateral branches produced (a) near the apex of a main hypha, and (b) further back.

Notes:

1. To prepare plates of *Saprolegnia* take a 5 mm diameter disc from a colony grown on 0.1 per cent malt agar and use this as an inoculum. Use a flamed cork borer to obtain the discs. For plates of *A. niger* and *G. candidum* streak spores on a sterile loop across a diameter of each plate.
2. A small square (2 x 2 cm) of a polypropylene membrane (Sterilin, Richmond, Surrey, England) can be laid carefully over the area of each fungus to be studied. The membrane is permeable to carbon dioxide and oxygen and facilitates high-power observation by reducing condensation on the front lens of the objective.
3. The following procedure is recommended for calibration of the eyepiece micrometer.
   - (i) Insert the eyepiece micrometer in the microscope to replace the normal eyepiece.
   - (ii) Mount the stage micrometer on the microscope stage and centre the scale using the low-power (x 10) objective.
   - (iii) Rotate the eyepiece micrometer and adjust the stage micrometer until the lines of both micrometers coincide at one side of the field of view. Count the spaces of each micrometer to a point where the lines coincide again. Repeat this procedure for each objective to be used.
   - (iv) Each space between the rulings of a stage micrometer normally equals 10 μm. The number of micrometres corresponding with each space of the ocular scale can now be calculated.
   - (v) A haemocytometer slide can be used as a stage micrometer.

Flowsheet 8. (continued)

---

4. If a small volume (10 ml) of medium is added to a plastic Petri dish the fungus inoculated can be observed, and hyphal growth measured, without removing the lid of the dish. Simply place the Petri dish (lid downwards) on the microscope stage and focus through the base of the dish. This technique eliminates any effects that might occur through the culture drying out or through placing a membrane over the hyphae.
5. Preparation of media. *0.1 per cent malt agar:* malt extract, 1.0 g; agar, 12 g; distilled water to 1 l.

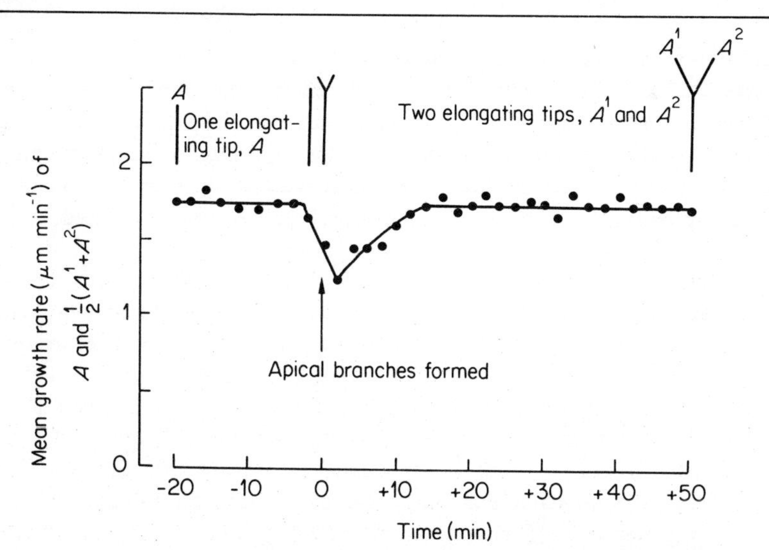

Fig. 9. Apical branching in *Geotrichum candidum.* The growth rate of the parent tip (A) and the mean growth rate of the apical branches ($A^1 + A^2/2$) are plotted. Note the reduction in extension-growth of the parent hypha just before branch initiation. Each point is the result of measurements on eight branching hyphae. Reprinted, with permission, from Trinci, *Transactions of the British Mycological Society,* **55** 17–28 (1970). Published by Cambridge University Press.

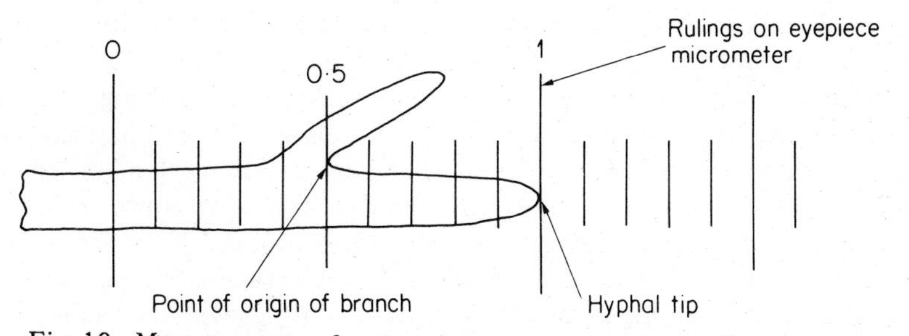

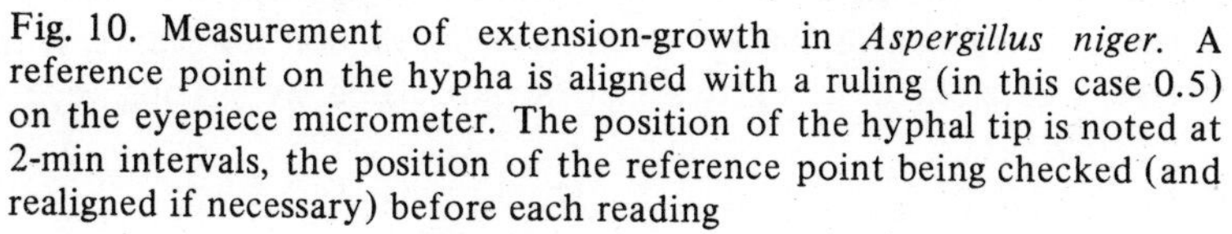

Fig. 10. Measurement of extension-growth in *Aspergillus niger.* A reference point on the hypha is aligned with a ruling (in this case 0.5) on the eyepiece micrometer. The position of the hyphal tip is noted at 2-min intervals, the position of the reference point being checked (and realigned if necessary) before each reading

region immediately adjacent to the tip. The formation of an apical or subapical branch is usually preceded by a slight reduction in the rate of extension-growth of the hypha (Fig. 9) but the technique outlined in Flowsheet 8 may not be sufficiently sensitive to detect this change. The newly-formed branch or branches soon reach the rate of extension-growth of the parent hypha.

2.2 INHIBITION OF HYPHAL GROWTH

Hyphae extend at a rate which is normally constant for a given set of conditions. The rate of extension-growth can be reduced by the application of various substances to the colony margin. In this exercise the effects of a fungicide, an inhibitor of protein synthesis, and an uncoupler of oxidative phosphorylation are studied. *Neurospora crassa* has been selected as the test organism on account of its broad, fast-growing hyphae. The experimental procedure is outlined in Flowsheet 9.

The effect of a fungicide (such as copper sulphate) on hyphal growth can also be estimated by incorporating different concentrations of copper sulphate into culture media and estimating the radial growth rate of colonies of fungi seeded on the media. Proceed as follows. Inoculate plates of Czapek's solution agar (see Flowsheet 23) centrally with conidia of *Trichoderma viride* and incubate at 25 °C for 5 days. Prepare a series of Czapek's solution agar plates containing 10^{-3}, 10^{-4}, 10^{-5}, and 10^{-6} M-copper sulphate. Inoculate each plate centrally with a 1 cm diameter disc cut from behind the colony margin of one of the cultures of *T. viride.* Incubate the plates at 25 °C and record the radial growth of each colony at daily intervals. The diameter (and hence the radius) of each colony is measured by inverting the dishes and placing a ruler flat against the glass or plastic base of each dish. Record the diameter to the nearest millimetre and make two measurements (at right angles to each other) for each colony. Do not forget to allow for the diameter of the disc used as inoculum. Graph radial growth rate of the colony (= leader hyphae) against time for each concentration of copper sulphate. Do your results indicate that some concentrations of copper sulphate stimulate rather than inhibit growth? As an extension of this experiment, study the different effects of copper sulphate concentration on the colony morphology of *T. viride.*

The effect of temperature on retardation and promotion of hyphal growth can be studied by the following method. Prepare plates of 0.1 per cent malt agar and inoculate each plate centrally with a 1 cm diameter disc cut from behind the colony margin of a 2-day-old colony of *Saprolegnia ferax* or *Achlya* sp. grown on 0.1 per cent malt agar. Incubate separate plates at 5, 10, 15, 20, 25, and 30 °C. Record the radial growth of each colony at daily intervals using the same technique as in the previous experiment. At the optimum temperature for growth the colony will probably reach the margin of the Petri dish after incubation for 3 days. If more measurements are required the plates can be inoculated by placing each disc

Flowsheet 9. Inhibition of hyphal growth

Inoculate plates of 0.1 per cent malt agar by transferring spores of *Neurospora crassa* to the centre of each plate with an inoculating needle. Cover the inoculum and most of the medium with a large square (5 x 5 cm) of sterile polypropylene film. Apply the film evenly and leave the plates at room temperature (20 °C) for 24 h.

↓

Cut each culture (membrane, colony, and medium) into eight sectors with a razor blade and leave for 0.5 h. Transfer one of the sectors onto a glass slide and locate one of the tips of the leader hyphae with the x 40 objective. Align the tip with one of the divisions of an eyepiece micrometer (see previous flowsheet) and measure the extension-growth at 2-min intervals for a period of at least 10 min. Graph your results.

↓

Mount another sector of the colony of *N. crassa* on a slide and measure the growth rate of one of the leader hyphae over a period of 6 min. The slide should be clamped securely on the microscope stage. Swing the objective aside and carefully lift up a corner of the membrane with fine forceps. Add one drop of a solution of copper sulphate (10 mM), cycloheximide (0.2 mM), or 2,4-dinitrophenol (5 mM). Lower the membrane again, locate the same hyphal tip and record the growth rate at 2-min intervals as before. Test each solution in turn on a different sector and also repeat the experiment with 1/10 and 1/100 dilutions of the solutions already tested. Record your results as percentage reduction in extension-growth against concentration of each growth inhibitor. Record how long the inhibitors take to act and if each hypha recovers.

Notes:
1. The chemicals and dilutions should be prepared in 0.1 per cent malt extract solution to be isotonic with the growth medium. As a control, add a drop of 0.1 per cent malt extract solution to the margin of one of the sectors and note any effect on the extension-growth of a selected hypha.
2. Each sector can only be used for one experimental treatment.

adjacent to the margin of the plate so giving the full diameter of the plate for the culture to spread. Measure from the margin of the disc to the furthest point of spread for each colony.

2.3. MECHANISM OF HYPHAL GROWTH

Hyphae grow apically. New cell wall material is synthesized and incorporated into the apical dome of the hypha. A positive hydrostatic pressure is maintained within the hypha, which results in the continuing expansion and forward movement of the apical region. This pressure can be readily demonstrated (Flowsheet 10). The treatment of hyphae of *Aspergillus niger* with a weak acid results in rupture at their weakest point (the apex) with an almost explosive violence. The hyphal contents are extruded very quickly in a long ribbon-like mass. This extrusion can be retarded for a considerable time by pretreatment of the hyphae with hypertonic solutions. So rapidly is water extracted from a hypha with hypertonic solutions that the hydrostatic pressure falls almost immediately and the hyphae do not burst in the presence of the acid. The hypertonic solution and acid can be added as one solution and the more concentrated the applied hypertonic solution the longer the delay

Flowsheet 10. Mechanism of extension-growth in hyphae

Inoculate plates of S agar with a diametric streak of spores from a malt agar slope culture of *Aspergillus niger.* Incubate plates for 3 days at 25 °C.

↓

Place a plate on the microscope stage and focus on the surface marginal hyphae with the x 10 or x 20 objective. Avoid hyphae at the margin of the plate. Raise the objective and add one drop of dilute acetic acid (0.5 per cent v/v) to the marginal hyphae beneath the objective and immediately lower the objective and refocus on the surface marginal hyphae. Note the time taken for most of the hyphae to burst.

↓

Repeat the procedure but this time add one drop of a 0.2 M-sucrose solution (prepared in 0.5 per cent v/v acetic acid) to a different portion of the colony margin. Try different concentrations of acetic acid and of sucrose in acetic acid and note the time taken for most of the hyphae to burst.

Notes:

1. The hyphae growing on the surface of the medium are first to respond to the treatments. They burst and their contents extrude as a ribbon-like mass. The hyphae just beneath the surface also burst but there is a time lag which increases with increasing depth. This is due to the time taken for the acetic acid to diffuse into the medium and also the resultant lower concentration of acid with increasing depth.
2. When hyphae below the surface of the medium burst the extruded material is in the form of a spherical mass which remains associated with the hyphal tip due to the resistance of the surrounding medium.
3. It is not advisable to carry out too many applications of acetic acid solutions on the same plate of *A. niger.* Hyphae exposed to the acetic acid vapour can acquire a tolerance such that they become less affected by the subsequent application of a drop of the dilute acid.
4. Not all the hyphal tips treated with acetic acid will burst, particularly when hyphae produce subapical or lateral branches which become leader hyphae. In such cases the rupture of one hyphal tip reduces the hydrostatic pressure in the associated branch(es) such that it does not burst.

before the hyphae burst. One explanation for this effect is that the more hypertonic solutions extract more water from the hyphae which in turn take longer to re-establish sufficient pressure to cause rupture.

This technique has been extended to estimate the osmotic equivalent of hyphae (Park and Robinson, 1966). Drops of acetic acid (0.5 per cent v/v) and of sucrose solutions (a range of concentrations prepared in 0.5 per cent v/v acetic acid) are added separately to different regions of the margin of a colony. The time taken for all the surface marginal hyphae to burst is noted for each application. At low sucrose concentrations the time to completion of bursting is the same as for acetic acid with no added sucrose but with increasing sucrose concentration the time

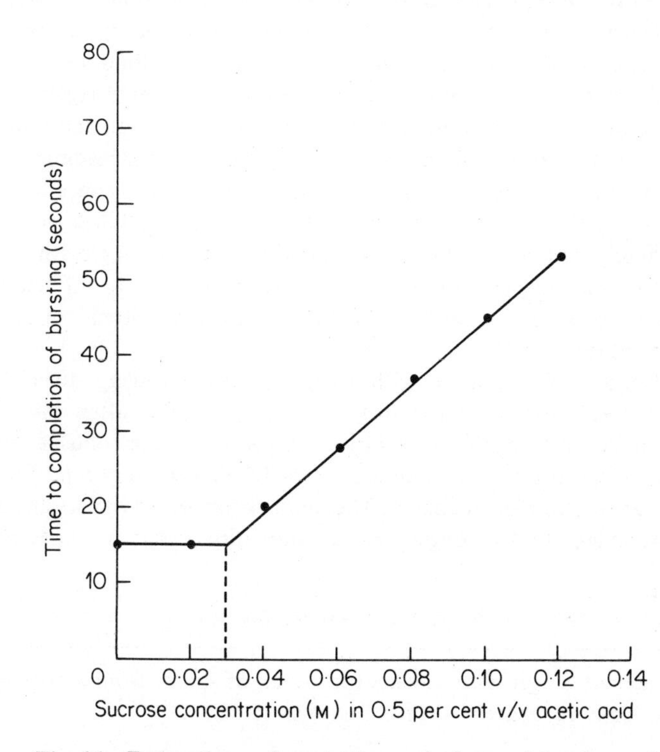

Fig. 11. Estimation of osmotic equivalent of hyphae of *Aspergillus niger*. Sucrose solutions in 0.5 per cent v/v acetic acid are applied to the colony margin and the time noted for all the hyphae in the field of view to burst. A x 20 objective is used. The time to completion of bursting begins to increase at a sucrose concentration of 0.04 M. The point of inflexion of the two lines indicates the osmotic equivalent of the hyphae. The graph shown is for a 2-day-old colony of *A. niger* grown on S agar. The value obtained for the osmotic equivalent is 0.03 M and each point is the mean of five determinations (Park and Robinson, 1966). Reproduced by permission of Oxford University Press

increases. When the time begins to increase, the applied solution is considered hypertonic to the contents of the hyphae. Typical results are shown in Fig. 11 and the point of inflexion of the two lines has been assumed to indicate the osmotic equivalent of the hyphal contents.

2.4 HYPHAL MORPHOGENESIS

The marginal or leader hyphae of colonies of most mycelial fungi grow radially away from the inoculum or colony centre. These hyphae are broad, and taper at the growing tip region. During growth in septate species the apical compartment or cell forms a series of subterminal compartments by the production of cross-walls and each of these compartments may produce a lateral branch. The apical compartment may on occasions produce lateral branches some distance behind the apex and also subapical branches which are produced just behind the tapered region of the tip. Branch formation occurs quite frequently during normal growth and development of the colony and is correlated with the nutrient status of the medium. Branching of the colony as a whole will be discussed later. This section is concerned with the morphogenesis of the terminal compartment of individual hyphae of *Aspergillus niger*, particularly the relationship between checking extension-growth for various times and the resultant growth pattern. A convenient way to influence hyphal morphogenesis is by the treatment of hyphae with hypotonic or hypertonic solutions (Flowsheet 11).

The application of a solution with a slightly greater tonicity than the apical compartment checks extension-growth for only a short time, often with no visible effect other than the formation of a slight constriction at the point of check. This constriction can persist and is sometimes a useful marker of the position of the hyphal apex at the time of treatment. The more concentrated hypertonic solutions arrest extension-growth for longer periods and initiate various types of branch

Flowsheet 11. Hyphal morphogenesis in *Aspergillus niger*

Inoculate plates of S agar with a diametric streak of spores from a malt agar slope culture of *A. niger.* Incubate plates for 3 days at 25 °C.

↓

Place drops of each of the following solutions on separate regions of the colony margin: 0.05, 0.10, 0.25, 0.50, and 1.0 M-sucrose. Add a drop of distilled water as a control.

Notes:

1. It is convenient to mark the base of the plate with a series of circles to denote the regions on the margin where the various solutions are to be applied.
2. Study hyphae on the surface of the agar and in the central region of the drop, not at the margin of the drop.
3. Record the time for which growth is checked and make notes on the resultant growth patterns.

formation. With many strains of *A. niger* the following sequence can be observed with increasing concentration of the applied hypertonic solution: a slight check to extension-growth, sometimes followed by the formation of a lateral branch in the vicinity of the point of check; a longer check to growth accompanied by a swelling of the hyphal tip, often by the later production of one or more subapical branches from the side of the swelling, growth does not usually continue along the original axis. This sequence is illustrated in Fig. 12.

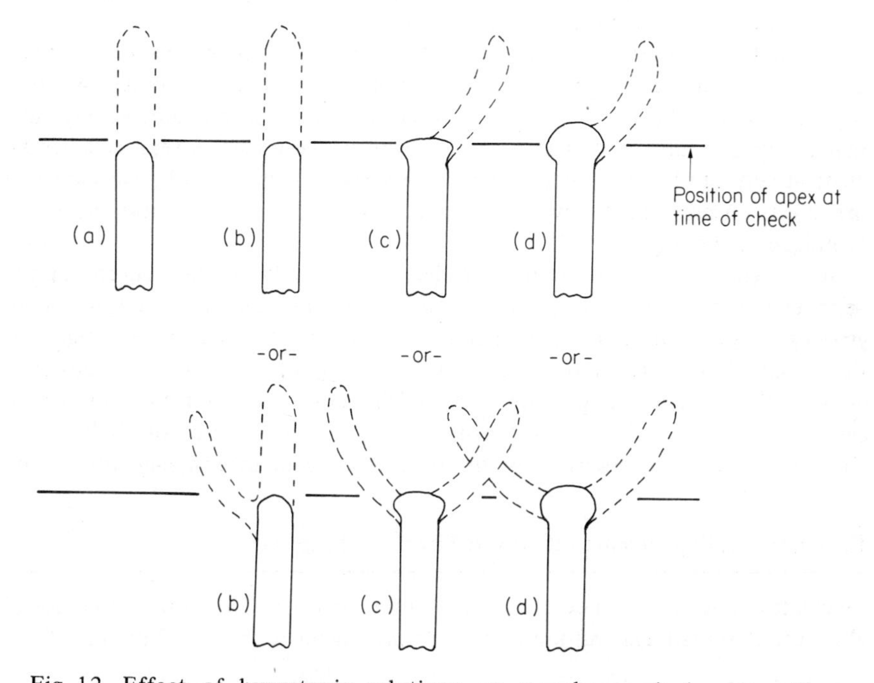

Fig. 12. Effect of hypertonic solutions on morphogenesis in *Aspergillus* niger. Three-day-old S agar cultures of *A. niger* are treated with sucrose solutions of the following molarities: (a) 0.2, (b) 0.4, (c) 1.0, and (d) 2.0 M. In (a) there is hardly any check to growth. In (b) there is a slight check with the apex flattening. Growth is resumed along the same axis but a lateral branch may form at the point of check. In (c) a longer check results and the apex becomes pestle-shaped. Extension-growth resumes by the formation of one or more subapical branches from the shoulder of the swollen apex. In (d) there is a very long check. The apex swells to become globose and, as in (c), growth resumes by the formation of one or more subapical branches (Park and Robinson, 1966). Reproduced by permission of Oxford University Press

2.5 HYPHAL SEVERANCE

Much information can be gained about the process of hyphal and colony growth by simple experiments in which hyphae are completely severed. It is recommended that a fast-growing species of *Saprolegnia* be studied by this means. Hyphae can be cut at various distances from the colony margin and the subsequent effects on extension-growth studied.

When hyphae of a *Saprolegnia* sp. (other species of Phycomycetes could be substituted) are severed at a point one centimetre or more behind the hyphal apices the hyphae cease growing immediately. If the severed region is examined it will be noted that the extent of visible damage is very slight and only a small portion of the hyphal contents appears to have extruded. The hyphal tip soon resumes extension-growth and this is preceded by the tip becoming swollen (Fig. 13). The experimental system is outlined in Flowsheet 12.

A variety of experiments can be carried out. Cuts can be made at varying distances behind the margin of different colonies and the effects noted. Another approach is to make a series of cuts at intervals of time on the same colony such that a hypha is just allowed to resume growth between cuts. Several cuts can be made at very short (say ten-second) intervals working towards the hyphal apex and the effects on subsequent growth resumption studied. Various modifications of this technique can be tried.

An extension of these experiments leads to a study of the concept of the peripheral growth zone which has already been discussed. In a fungal colony growing on a solid medium, growth is mainly restricted to the peripheral region of the colony. Numerous questions come to mind. To what extent is the observed growth at the colony margin due to materials translocated from the older (more central) regions of the colony? What portion of cytoplasm of each hypha contributes to and is essential for the observed rate of extension-growth of the

Flowsheet 12. Experiments with severed hyphae of *Saprolegnia* sp.

Inoculate plates of 0.1 per cent malt agar with rectangular (5 cm x 0.5 cm) strips of a 0.1 per cent malt agar culture of *Saprolegnia.* Incubate for 1 or 2 days at 25 °C.

↓

Note the rate of extension-growth of one or more of the marginal hyphae growing on the surface of the medium. Cut well behind (1 cm or more) the colony margin with a razor blade, severing all of the marginal hyphae with a firm cut through the medium. Note the behaviour of the severed hyphae, particularly any changes in morphology and the time taken for these changes to occur. Note the time taken for growth to recommence. Is the former rate of extension-growth established?

↓

Make an oblique cut in relation to the colony margin. This will result in some hyphae being severed close to the hyphal apex and others being severed much further back. Note the reaction of the severed hyphae, particularly those which have been cut close to their apices. What is the minimum length of hypha (a) which can survive (continue growth), and (b) which can continue growth at the same rate as before cutting?

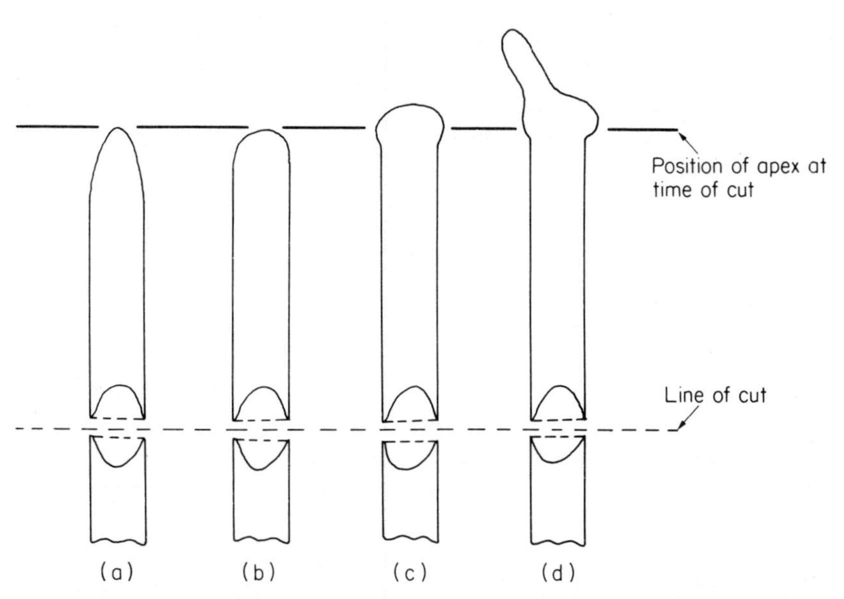

Fig. 13. Hyphal severance. Hyphae of *Saprolegnia* are cut with a razor blade at a point 1 cm behind the colony margin. Immediately after cutting (a) the hyphae cease growth. Later, (b) and (c), the apex becomes rounded and globose. Finally (d), growth resumes from one side of the swollen tip. The whole sequence takes just a few minutes. (Robinson, unpublished data)

marginal hypha? It is possible to answer partially these questions by cutting hyphae at various distances from their apices and observing the resultant rate of extension-growth when growth resumes after the initial shock of cutting. All the severed hyphae are able to grow independently of the colony provided they are not cut so close to the apex as to be killed. It has been assumed that the length of hypha which, after cutting, is just able to grow at its former rate is equivalent to the width of the peripheral growth zone of the fungus.

2.6 TROPIC RESPONSES OF HYPHAE TO NUTRIENTS

The centrifugal growth of hyphae from an inoculum is an obvious example of an orientated growth response. As will emerge later in Section 3.3, it is an exceedingly difficult problem to decide if the hyphae are growing away from some factor (as yet unidentified) or growing in response to a diffusion gradient of one or more nutrients at the margin of the colony. Obviously the nutrient concentration of a solid medium will be greater outside the perimeter of a developing colony and it seems reasonable to consider the possibility that hyphae may respond to this concentration gradient of nutrients.

It is not yet possible to demonstrate chemotropic responses to nutrients by hyphae of many species of fungi but these tropisms can be induced in some aquatic Phycomycetes by quite straightforward methods (Fischer and Werner, 1955;

Flowsheet 13. Tropic responses of hyphae to nutrients

Inoculate plates of 0.02 per cent malt agar with small discs cut from cultures of species of *Saprolegnia* or *Achlya.* Inoculate the plates at the margin and incubate for 1 or 2 days at 25 °C.

↓

Prepare a plate of 5.0 per cent malt agar. Remove a disc (5 mm diameter) with a sterile cork borer and place the disc just behind the margin of the colony. Observe the behaviour of the neighbouring hyphae over a period of several hours.

↓

Repeat the experiment using discs of casein hydrolysate ($1\ g\ l^{-1}$) incorporated into agar and also individual amino acids such as L-cysteine ($10\ g\ l^{-1}$) in agar.

Notes:
1. It is recommended that 20 g agar l^{-1} is incorporated into all media (even the test media from which discs are cut). The plates on which the test fungi are grown should be poured thinly (approximately 10 ml medium/plate). *0.02 per cent malt agar:* malt extract, 0.2 g; agar, 20 g; distilled water to 1 l.
2. When placing the test discs behind the colony margin it is useful to hold the plate up to the light so that the colony margin is clearly visible. The test disc can then be picked up with a sterile dissecting needle and placed approximately 1 mm behind the margin of the colony. In practice it is convenient to place the disc a little further behind the margin and to push the disc radially to the required position. In this way the radial symmetry of the hyphae is not disturbed.
3. The assay system can be modified by inoculating plates of 0.02 per cent malt agar centrally with a rectangular (5 cm x 0.5 cm) strip cut from behind the margin of a culture grown on 0.1 per cent malt agar (20 g agar l^{-1}). After incubation of this culture for one day at 25 °C test discs can be placed overlapping the colony margin and the response of neighbouring hyphae measured at time intervals of 0.5 h or so. With this method the hyphae (before treatment) grow parallel to each other and perpendicular to the edge of the rectangular inoculum strip. The effect of various test discs can be estimated by measuring the distance along the margin over which hyphae respond (Fig. 15).
4. Tropisms to individual amino acids are best observed when the test fungus is grown on the following medium: casein hydrolysate, 0.2 g; agar, 12 g; distilled water to 1 l.

Musgrave *et al.*, 1977). These techniques are outlined in Flowsheet 13 and are suitable for many species of *Saprolegnia* and *Achlya.* The procedures progress from the demonstration of a tropic response to malt agar medium to the determination of the specific components responsible for the effect. For these experiments it is essential to grow the test fungi on a very weak medium otherwise a good response will not occur.

With many species of *Saprolegnia* or *Achlya* a tropic response can be detected as early as ten minutes after treatment. Variations are possible on these experiments and it is intended that the assay systems described could be used to test numerous compounds for possible tropic effects on hyphae. It will be noted that when a disc of 5.0 per cent malt agar is placed just behind the margin of a colony of the test fungus some of the hyphae in front of the disc will turn through an angle of 180° and grow towards the disc, in the direction of the centre of the colony. Marginal hyphae on either side of the disc turn and grow towards the disc and lateral branch production is stimulated on many of the marginal hyphae adjacent to the disc. These lateral branches frequently emerge from the side of the hypha facing the disc and grow towards the disc. The tropisms of the leader hyphae and of their lateral branches are readily observed in this system as the test fungi produce well-separated marginal hyphae when grown on a weak medium and few lateral branches are normally produced. Some of these tropic responses are shown in Fig. 14.

Having demonstrated that hyphae can exhibit a positive chemotropism to a higher concentration of the same medium on which they are already growing, it is of interest to determine the ingredient or ingredients of the medium which are responsible. It will soon become apparent that it is the amino acid components of the medium which are active in eliciting the response. If the experimentation is taken further it can be shown that the tropic effect is more marked with mixtures of amino acids than with single amino acids. In fact, cysteine is usually the only amino acid which can induce a response when assayed individually and it is essential to use L-cysteine as the D-isomer is inactive. Also, relatively high concentrations of L-cysteine are required to give an effect, in contrast to the low concentrations of L-amino acids which are effective as a mixture.

Numerous substances can be tested for chemotropic activity quite simply by an alternative method in which small quantities of solid or liquid material are placed behind the colony margin. It is of interest that Wortmann (1887) tried flies' legs!

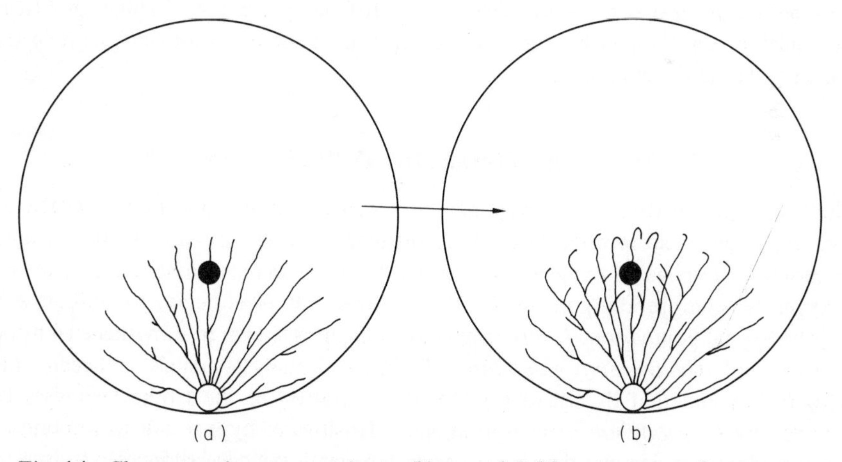

Fig. 14. Chemotropism to nutrients. Plates of 0.02 per cent malt agar are inoculated at the margin with *Saprolegnia.* After 1 day's growth a disc of 5.0 per cent malt agar or casein hydrolysate (0.1 per cent) in agar is placed just behind the colony margin (a). After one hour or so (b) some of the leader hyphae have turned towards the disc and lateral branch formation is stimulated in the vicinity of the disc (After Fischer and Werner, 1955). Reproduced by permission of Walter de Gruyter & Co.

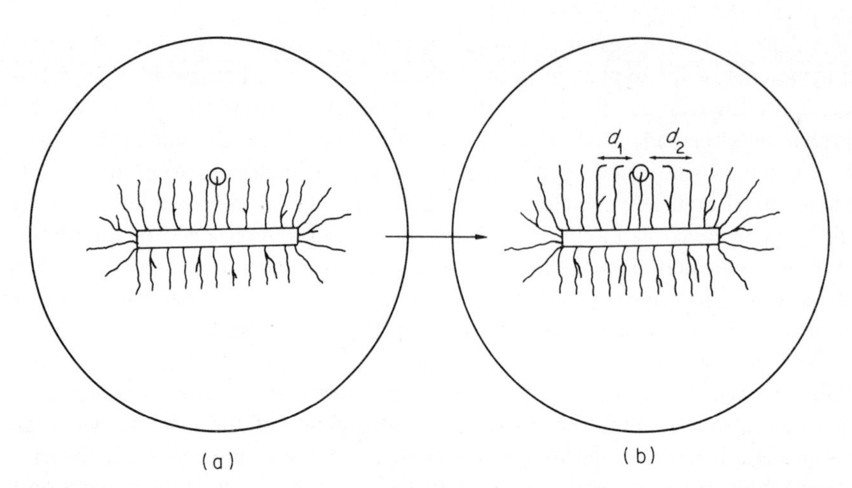

Fig. 15. Assay for chemotropic responses in *Saprolegnia.* Colonies of *Saprolegnia* are grown from rectangular slabs of a plate culture. The plate is incubated for 1 day and a test disc placed on the margin of a colony (a). After a selected time (b) the distance over which hyphae are affected is measured ($d_1 + d_2/2$) . (Robinson, unpublished data)

A word of caution. Not all fungi respond to concentrated media or amino acids as do the recommended test genera. Neither do all species of *Achlya* and *Saprolegnia* respond. These tropic responses to nutrients have not been generally detected throughout the fungal kingdom although it may be that suitable assay systems have not been developed to demonstrate them. Finally, these experiments by no means prove that the observed centrifugal pattern of hyphal growth in colonies is conditioned by the concentration gradient of various nutrients in the vicinity of the marginal hyphae.

2.7 TROPIC RESPONSES OF HYPHAE TO OXYGEN

Just as a germinating spore responds to a concentration gradient of oxygen so mature hyphae can respond in a similar manner. In order to demonstrate a tropic response of hyphae to oxygen it is essential to create an environment such that the oxygen tension prevailing in the vicinity of the test hyphae begins to limit growth of the hyphal population. A very simple experiment is shown in Flowsheet 14 using cultures of *Geotrichum candidum.* There are many parallels between this experiment and that described for the demonstration of the tropic responses to oxygen by *G. candidum* arthrospores. Low densities of hyphae fail to respond as oxygen does not become limiting for growth beneath the coverslip which is used to restrict gaseous diffusion from the atmosphere.

Several responses become apparent at the higher nutrient levels. Higher hyphal densities occur at higher nutrient concentrations and shortly after application of the coverslip many of the original marginal hyphae cease growing and this results in a reduced hyphal density. Beneath the centre of the coverslip very few of the hyphae form lateral branches from the cells laid down by each advancing apical cell. The new cells laid down are sometimes twice as long as cells produced outside the area of the coverslip. The leader hyphae which ceased growth after application of the coverslip become vacuolate, the vacuoles extending to the hyphal apices.

The hyphae growing just inside the lateral margin of the coverslip and parallel to the sides respond differently. Some of the hyphae turn towards the margin of the coverslip and this is presumed to be a positive chemotropic response to oxygen. These hyphae produce lateral branches from their non-apical cells and the branches are often produced from the side of the cell facing the coverslip margin. The branches tend to grow towards the coverslip margin. These responses are illustrated in Fig. 16.

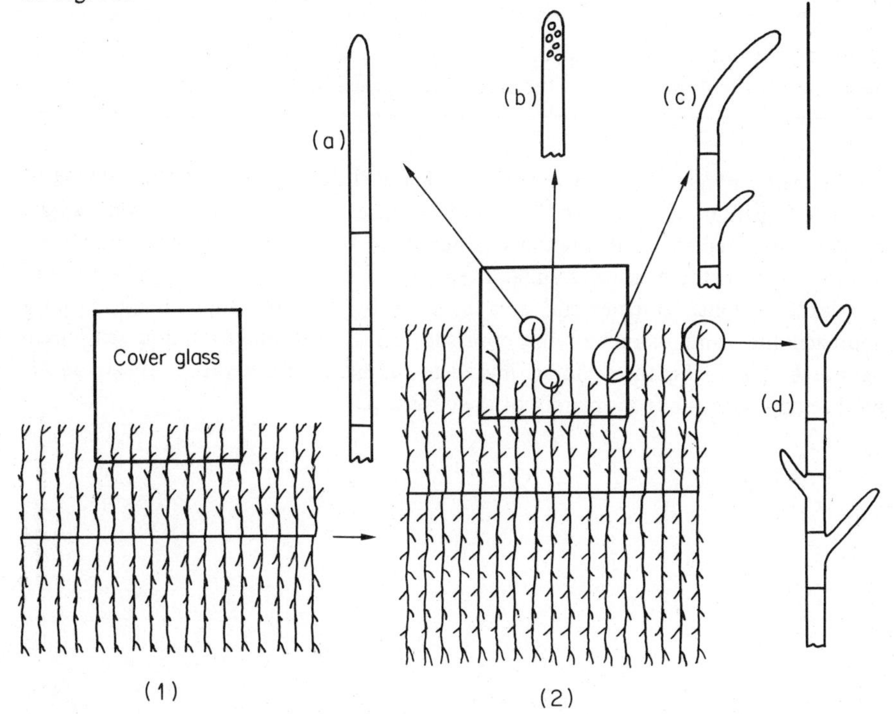

Fig. 16. Effect of reduced oxygen tension on cultures of *Geotrichum candidum*. A coverslip is placed over the margin of a 3-day-old colony of *G. candidum* on 1 per cent malt agar (1). One day later (2) the following effects will have occurred. (a) In the central region of the coverslip the hyphal density decreases and the hyphae which continue extension-growth do not produce lateral branches. The cells which form are at least twice as long as cells outside the coverslip. (b) The hyphae which cease growth become vacuolate at the apex (normally a non-vacuolate region). (c) The hyphae near the margin of the coverslip turn towards the margin and lateral branches frequently arise from the side of the cell nearer the margin. (d) A 'normal' hypha (outside the coverslip). The terminal cell may branch subapically and a lateral branch usually forms at the distal end of each cell which is produced. Reproduced, with permission, from Robinson and Griffith, *Transactions of the British Mycological Society*, **68** 311–314 (1977). Published by Cambridge University Press.

Flowsheet 14. Tropic responses of hyphae to oxygen

Inoculate plates of 0.05, 0.10, and 1.0 per cent malt agar with a diametric streak of arthrospores of *Geotrichum candidum.* Incubate plates for 3 days at 25 °C.

↓

Place a clean (sterile) coverslip over a portion of the margin of the colony on each plate such that the tips of the marginal hyphae are approximately one-quarter of the way across the coverslip and the sides of the coverslip are aligned parallel to the hyphae.

↓

Incubate the dishes at 25 °C and examine at 3 and 24 h. Make notes on the behaviour of the hyphae within the area of the coverslip.

The leader hyphae beneath the central area of the coverslip continue growing to the front margin of the coverslip; these hyphae adjust to the reduced oxygen tension by a reduction in numbers rather than by a reduced rate of extension-growth. In fact, these leader hyphae keep pace in terms of extension-growth with the leader hyphae outside the area covered by the coverslip. As the hyphae approach the front margin of the coverslip lateral branches form and cell length decreases. These responses are attributed to the increase in oxygen tension as the hyphae approach the front margin of the coverslip.

3
The Developing Colony

When a solid culture medium is inoculated with fungus spores hyphae develop in several planes and produce branches. As the hyphae extend and more branches form the colony becomes visible to the unaided eye. At this stage of development, usually after one or two days, the colony is circular in outline and the margin comprises actively growing hyphae which are almost parallel to each other. As already described, the colony continues to expand at a constant radial growth rate provided growth is not limited by lack of nutrients or by some other factor.

In liquid culture there is less tendency for colonies to occur in the form just described. The hyphae produced by the germinating spores become tangled and a diffuse mycelium develops throughout the medium with growth occurring simultaneously in all regions of the medium. The gradients in nutrient concentration which occur in the vicinity of a colony grown on a solid medium are much reduced in liquid culture, particularly in shake-flask cultures, as are the gradients in concentration of various self-produced metabolites. Included in these self-produced metabolites are 'staling substances' or autotoxic metabolites and these are the subject of a later discussion. The presence of gradients in both nutrient and autotoxic metabolite concentration in plate cultures probably accounts for the greater degree of hyphal organization and differentiation encountered in this method of culture which is the subject of the first experiment in this chapter.

The morphology of fungal colonies is affected by environmental and genetic factors. The effect of environment can be readily demonstrated by growing a fungus on media which contain high and low concentrations of nutrients. Fewer branches form on media low in nutrient concentration. Experiments by Buller (1933) emphasize the importance of genetic control; in *Coprinus fimetarius* the angle between a lateral branch and its parent hypha was 70°–90° in monokaryons and 20°–40° in the dikaryon. The genetical control of branching is also illustrated by the colonial mutants of *Neurospora* in which some mutants produce lateral branches which are close together and in which the growth habit is dichotomous rather than monopodial.

Since the branching pattern of a colony has a large bearing on its morphological appearance it is not surprising that the interrelationships of branches have been studied by several workers. Butler (1961) has described experiments in which the extension-growth rates of main hyphae, primary lateral branches, and secondary lateral branches were recorded for colonies of *Coprinus disseminatus*. In most instances the rates of extension of the main hyphae, primary laterals, and secondary laterals were constant in the ratio 100:66:18. Although the main hyphae were usually wider than the branch hyphae it was found that when main and primary branch hyphae were of the same diameter the main hyphae still grew faster. Obviously the growth of a hypha, whether a main or lateral branch hypha, is related to the origin of the hypha as well as its size. The situation has been compared with that in higher green plants where growth of lateral branches is influenced by the main growth axis. In higher plants the phenomenon is termed 'apical dominance' or 'correlative inhibition' and different hypotheses have been proposed to account for the phenomenon. The 'diversion theory' proposed by Went (1936) to account for correlative inhibition may have some relevance to the situation in fungi. The faster growing (and usually wider) main hyphae could divert, or compete more readily for, one or more metabolites which affect or even regulate hyphal growth. Some evidence for this idea comes from an observation made by Butler (1961). She found that when a lateral branch ceased growth the adjacent laterals increased in growth rate. This effect suggests some type of internal competition between branches.

3.1 MORPHOLOGY OF THE DEVELOPING COLONY

While fungi of different species vary markedly as regards their overall morphology there are several general features which are common to colonies of most mycelial fungi (Park, 1961). This exercise is purely morphological and it will soon become apparent that the morphological pattern observed is remarkably constant for a particular species growing under defined conditions. The exercise poses several questions which, although seemingly straightforward, have yet to be unequivocally answered.

The fungus selected for this exercise is *Fusarium oxysporum* but many other fungi could be studied in a similar manner. Preparation is simple; plates of S agar are inoculated centrally with spores of *F. oxysporum* and incubated for four days at 25 °C. Examine the colonies directly with a microscope (x 20 objective, if possible) commencing at the margin of the colony and working towards the centre. Several zones can be distinguished and the relative sizes of these are illustrated in Fig. 17; the zones are lettered (a)–(g).

The first zone (a) is an obvious one and consists of broad, straight-growing marginal or leader hyphae. These hyphae are growing radially away from the colony centre and not much branching is apparent. In the next zone (b), moving inwards slightly from the periphery of the colony, much more branching is apparent. The branches are not as wide as the marginal hyphae. Note the orientation of these branches; they turn towards 'open' (free of hyphae) areas of medium and then grow radially towards the colony margin. Moving further inwards, zone (c), the branches are profuse. When the colony is viewed from above it will be seen that areas of medium are completely sealed off by these branches. Lateral branches are formed which grow into these areas. These branches are very narrow and grow in almost any direction, seldom do they grow straight towards the colony margin. Hyphal loops and coils are present, the hyphae having the appearance of coils of rope.

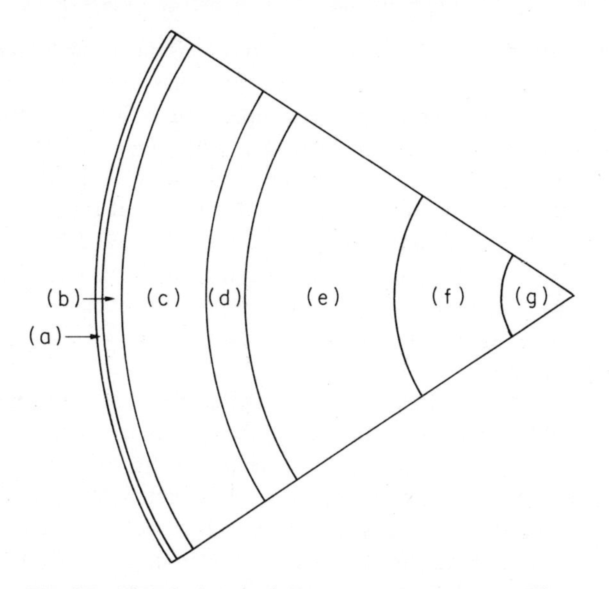

Fig. 17. Morphology of *Fusarium oxysporum*. The diagram illustrates a sector from a 4-day-old colony of *F. oxysporum* on S agar. The relative sizes of the zones referred to in the text are shown. The actual widths of these zones are as follows: (a) 200 μm, (b) 500 μm, (c) 2.3 mm, (d) 1 mm, (e) 4 mm, (f) 3 mm, and (g) 2 mm. These values will vary from strain to strain of the fungus. (Drawn from data of Park, 1961)

Nearer the centre of the colony, zone (d), some of the shorter lateral branches produce microconidia (small, single-celled spores) successively to form small groups of spores. Still further in, zone (e), autolysis (dissolution of the cell contents) is seen in some of the wide primary (main) hyphae and also in some of the straight lateral branches. In some strains of *F. oxysporum* terminal and intercalary chlamydospores (thick-walled spores) develop in the narrow, meandering hyphae of the next zone (f). Near the colony centre, zone (g), autolysis of all types of hyphae occurs and there is a low density of living hyphae. There are many chlamydospores present and groups of microconidia which do not appear to be associated with any living hyphae. This latter observation is the subject of a later experiment.

These observations may vary somewhat from strain to strain of *F. oxysporum*, particularly with reference to the types of spore produced. Not all strains will produce the three types of spore known for this species. Two types, the microconidium and the chlamydospore, have already been mentioned. The third type is the macroconidium which is larger than the microconidium, contains more than one cell, and is usually sickle- or banana-shaped.

Observe the morphological changes which have been outlined and consider the following questions. What shape does the hyphal apex have? Where do septa begin to develop? Can you observe septa developing; if so, how long does a septum take to develop? Where do branches develop? What is the behaviour of the lateral branches nearest the tip? Does each of these branches form a hypha just the same as that from which it was produced? How do the branches produced further back from the margin behave? Notice their orientation. Does it differ from that of hyphae nearer the margin? Can you see any evidence of hyphae fusing with each other? What significance might hyphal fusions have? What types of spore can you find? Where, and how, are they formed (a) in the colony, and (b) in relation to the hyphae? How much of the fungus near the centre of the colony is still alive? Stain with lactophenol + acid fuchsin and observe the distribution of protoplasm. If hyphae are undergoing autolysis or loss of contents, do all the cells in a hypha behave in the same way?

These are just a few examples of questions that might occur during examination of the colony of *F. oxysporum*. Many of these questions would be relevant to the examination of a colony of almost any mycelial fungus. Think of other questions needing an answer or situations demanding an explanation. Make notes and drawings to illustrate these queries, and any answers which suggest themselves. The answers themselves may promote further questions, e.g. as to how a particular pattern arises or is determined. How can it be altered? What experiments could be designed that might show which of the possible causes are operative?

3.2 EFFECT OF NUTRITION ON COLONY MORPHOLOGY

The morphology of a fungal colony can vary considerably according to the medium on which it is grown. A means to demonstrate this variation is outlined in Flowsheet 15. Colonies of *Geotrichum candidum* are grown on a range of media in which the initial concentration of glucose differs. Two main points emerge from this experiment. First of all it will be noted that the diameters of the colonies do not vary very much with time throughout the range of media tested. Second, it will soon emerge from merely holding the plates to the light that the main effect of increasing the glucose concentration is an increase in the hyphal density of the colony. This experiment demonstrates that in media where growth (as defined by increase in dry weight) is limited by the concentration of glucose in the medium,

Flowsheet 15. Effect of nutrition on colony morphology in *Geotrichum candidum*

Inoculate plates of the following media centrally with *G. candidum*: S agar containing 5, 10, 20, 50, and 100 mg glucose l^{-1}. Incubate plates for 3 days at 25 °C and examine.

Notes:
A convenient way of making and storing media is to prepare 200 ml aliquots of each medium in a medical flat bottle. The ingredients appropriate for a 200 ml quantity of medium are added to the bottle, water added to give a volume of 200 ml and the medium autoclaved for 15 min at 15 p.s.i. The medium can then be poured into plates or stored. If stored, the medium can be melted by autoclaving for 5 min at 10 p.s.i. Always shake the medium gently before pouring as some ingredients tend to layer.

the maximum radial growth rate (extension-growth) of the hyphae is achieved at very low glucose concentrations. This is true for many other fungi and Fig. 18 illustrates results for colonies of *Mucor hiemalis* and *Aspergillus nidulans*. The fact that fungi can spread so rapidly on a medium low in glucose or some other growth-limiting carbon source is obviously advantageous in natural habitats since even substrates low in nutrients would be rapidly colonized.

The main differences in colony morphology can be seen with the unaided eye but with the microscope other details emerge and some of the main changes in morphology are illustrated in Fig. 19. At the very low initial glucose concentrations (5 and 10 mg l^{-1}) there are very few marginal hyphae and a lateral branch usually develops from each cell produced by the apical cell of each marginal hypha. These lateral branches are well developed as they grow to form hyphae which are several cells long. Ultimately arthrospores may form from the terminal cell of each branch. Each leader hypha and its concomitant branches resembles a fir tree in branching habit. With a higher initial glucose concentration (20 mg l^{-1}) more leader hyphae are apparent and whereas the cells produced by the apical cell of each leader hypha form branches, these branches do not become as long as those produced at the lower glucose concentrations. As at the lower glucose concentrations, the terminal cell of each of the more proximal (i.e. nearer the colony centre) branches forms arthrospores. Arthrospores are formed by septation of the terminal cell and this is preceded by cessation of extension-growth of the cell. At the highest initial glucose concentrations (50 and 100 mg l^{-1}) there is a further increase in the density of the marginal hyphae. Many of the cells produced by each apical cell produce a lateral branch but these branches are relatively short and soon form arthrospores by division of the apical cell of each branch. The most striking change in growth pattern on the richer media is the capacity for the terminal cell of each leader hypha to branch subapically and so maintain a high hyphal density at the margin of the colony. It is clear that at the lower glucose concentrations the apical cells of the marginal hyphae rarely branch and even though the lateral branches are well developed there are relatively large areas of the medium uncolonized within the colony perimeter. At the higher glucose concentrations, due to the repeated branching of the apical cells, there are only very small areas of medium which are uncolonized. This results in a good coverage of the medium and is presumably reflected in the reduction in size of the lateral branches, one of which normally arises from the distal end of each cell produced by the apical cell of each leader hypha.

These observations can be extended in many ways. Are the apical and non-apical cells formed on the different media of different dimensions? Does each non-apical cell form a lateral branch on each medium? Is the radial pattern of growth of leader hyphae and lateral branches prominent on each medium? Many other questions will spring to mind.

In this experiment hyphal density is limited by the initial glucose concentration in the medium. In an earlier experiment (Section 2.7) hyphal density was assumed to be limited by the oxygen concentration in the medium. Compare notes on the two systems.

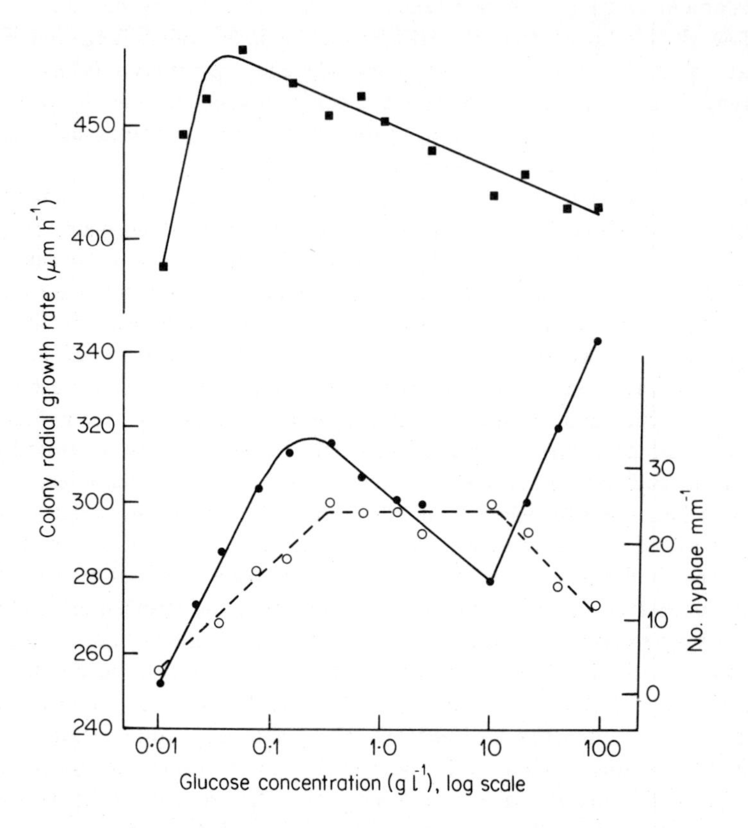

Fig. 18. Effect of glucose concentration on radial growth rate of *Mucor hiemalis* (■) and *Aspergillus nidulans* (●), and on hyphal density in *A. nidulans* (○). Both species grow rapidly even at the lowest glucose concentration (0.009 g l^{-1}) and colony growth is glucose-limited at low glucose concentrations. The radial growth rate of *A. nidulans* colonies is linear when plotted against the log of the initial glucose concentration over the range 0.009 g to 0.15 g l^{-1}. The optimum glucose concentration for maximum colony growth rate is 0.075 g l^{-1} for *M. hiemalis* and 0.2 g l^{-1} for *A. nidulans* (over the lower range of glucose concentrations studied). At glucose concentrations greater than the optimum there is a decrease in the growth rate in both species. At glucose concentrations above 10.24 g l^{-1} there is a second acceleration phase in the growth rate of colonies of *A. nidulans.*

The decline in radial growth rate in colonies of *A. nidulans* at higher glucose concentrations is correlated with attainment of maximum hyphal density. The acceleration in growth which commences between 10.24 and 20.48 g l^{-1} is correlated with a decrease in hyphal density (Trinci, 1969). Reproduced by permission of Cambridge University Press

The effect of nutrient concentration on colony morphology can also be demonstrated with species of *Achlya* or *Saprolegnia*. Prepare a series of plates of malt agar containing 0.05, 0.1, 0.25, 0.5, and 1.0 per cent malt extract respectively. Inoculate each plate centrally with a 1 cm diameter disc cut from just behind the margin of a 2-day-old colony grown on 0.1 per cent malt agar. Incubate the cultures

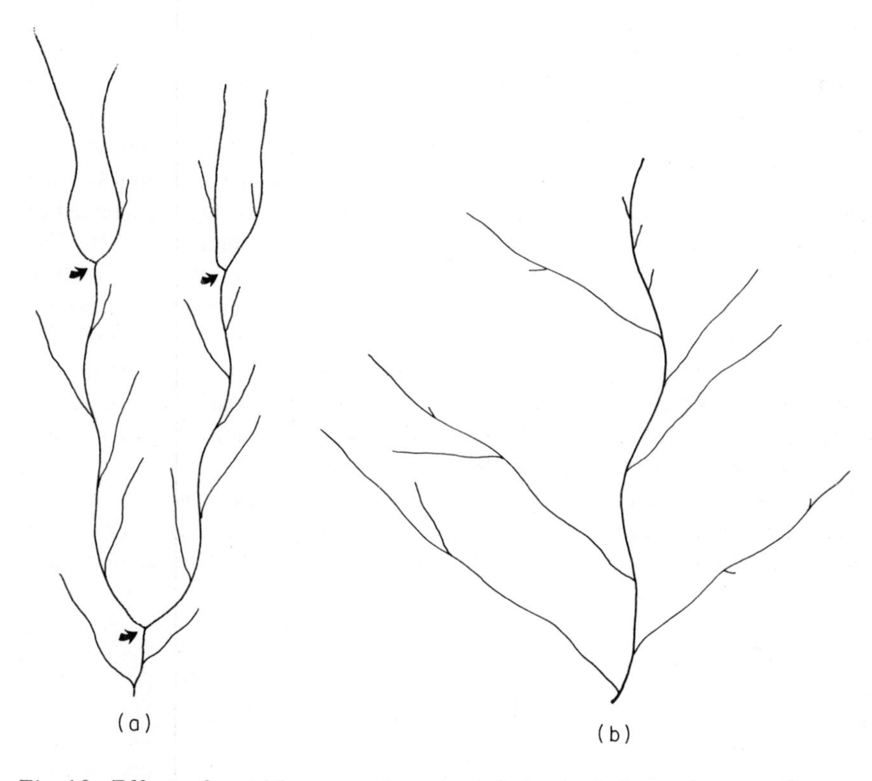

Fig. 19. Effect of nutrition on colony morphology of *Geotrichum candidum*. Growth patterns of hyphae cultured on media containing (a) 100, and (b) 10 mg glucose l^{-1}. The apical cells of the leader hyphae branch dichotomously in (a) and the three dichotomies which have occurred are arrowed. In (b) the apical cells do not branch but the lateral branches formed from other cells are better developed that in (a). (Robinson, unpublished data)

for two days at 25 °C and, after removing the lids, examine the margin of each colony with the aid of a microscope with a x 10 objective. Investigate how the hyphal density varies with medium concentration. Determine the number of lateral branches which are formed per unit length of leader hypha on each medium. Do lateral branches ever emerge very close to the apex of a leader hypha or is there a definite terminal length of hypha which does not branch?

3.3 MECHANISM OF CENTRIFUGAL GROWTH AND PATTERN IN THE DEVELOPING COLONY

Attention has already been drawn in Section 2.6 to the centrifugal growth of hyphae from an inoculum or the centre of the colony. Also, in Section 3.1, the sequence of changes in morphology from the margin to the centre of a colony was described and a characteristic pattern was found to exist for a particular species. What determines all these aspects of development? Consider first the centrifugal growth of hyphae. There are two main schools of thought and perhaps the most popular explanation is that fungi as they grow produce metabolites which influence hyphal growth and development. The autotropic factors referred to in Section 1.4 are examples of such metabolites. These are very relevant to any discussion on centrifugal growth since most colonies originate from a spore inoculum and the negative autotropism exhibited by germinating spores of most species examined to date would initiate the observed centrifugal growth pattern (Fig. 20). The continued production of such metabolites by the developing colony could result in the overall centrifugal growth pattern if it were assumed that these metabolites were present at a higher concentration in the older, more central regions of each colony. Before proceeding further, it is important to be absolutely clear on one point. There is no convincing demonstration that these metabolites actually exist. All attempts to isolate them have failed and this is not surprising if they are in fact as labile as they are postulated to be (Müller and Jaffe, 1965).

All the hypotheses based on metabolites to which germ-tubes and mature hyphae respond in a negative chemotropic manner owe their origin to work in which fungi appear to 'grow away from themselves'. Numerous researchers have

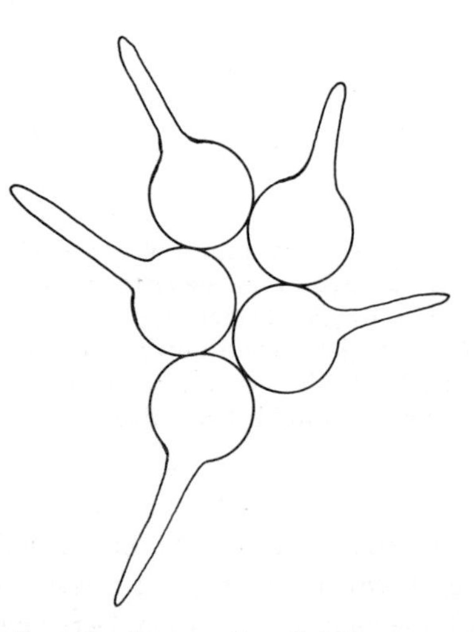

Fig. 20. Centrifugal growth of *Rhizopus stolonifer.* Spores of this species exhibit negative autotropism such that centrifugal growth of the hyphae is favoured in clumps of spores (Stadler, 1952). Reproduced by permission of The Wistar Press

shown that hyphae turn towards media containing a lower density of fungal material and away from media containing a higher density. They concluded that fungi produce substances which induce negative chemotropic responses in germinating spores and mature hyphae (Clark, 1902; Fulton, 1906; Graves, 1916; Stadler, 1952). It is important to consider the other point of view (although this has rarely been done) that the hyphae are exhibiting a positive tropism to factors, possibly nutrients, present at greater concentrations outside the colony margin. However, apart from the evidence that certain of the Phycomycetes respond to amino acids by positive chemotropism (Section 2.6) there is no convincing evidence as yet that fungi in general exhibit tropic responses to nutrients.

Another difficulty arises in so much as these two hypotheses become confounded in many experiments which are intended to demonstrate or prove one of them. A simple example will serve to illustrate this. In the preceding exercise (Section 3.2) the effect of nutrition on colony morphology was studied. It may have been noted that the centrifugal growth pattern, in particular the tropic responses of the lateral branches, was more marked at the higher glucose concentrations. It could be assumed that this is caused by increased production of metabolites (from which the hyphae turn) as a result of a higher mycelial density and metabolic activity at the higher nutrient levels. It could equally be due to a more marked response by the hyphae to particular nutrients at the colony periphery. This could result from the formation of a steeper concentration gradient of one or more nutrients essential for growth than would occur in cultures grown in media containing less glucose.

The situation is further complicated by the concept of 'staling substances' or autotoxic, growth-retarding metabolites (often labile) produced by fungus colonies. The hypothetical metabolites to which hyphae are supposedly negatively chemotropic have been called 'staling substances' (Stadler, 1952) although there is no evidence that these metabolites retard growth. Staling substances have also been invoked in discussions of the morphological pattern exhibited by most colonies of mycelial fungi from the colony margin to the colony centre (Section 3.1). The suggestion is based on the idea that not only do staling substances have an inhibitory effect on fungal growth but also that they induce morphogenic effects within the culture. As an area of a colony ages so the concentration of staling substances is assumed to increase and may result in the pattern of morphology already described (Section 3.1). The simplest hypothesis would be that colony morphogenesis is due to the accumulation of a single staling substance in the medium. Once again it is necessary to be cautious when interpreting many of the experiments which claim to demonstrate staling metabolites. As an area of a colony becomes older so will the nutrient concentration decrease in the medium associated with this portion of the colony. The concept of a staling substance or substances has to be considered in the context of this nutrient depletion. The morphogenetic changes that constitute staling can be correlated with the reduced nutrient status of the culture medium and it would be difficult to argue the precise trigger of differentiation – staling substance(s) or particular nutrients at a growth-limiting concentration.

Perhaps the most striking aspect of the morphological pattern exhibited by mycelial fungi is the pattern of branching. Many factors affect branching in fungi (Robertson, 1965) and there is considerable support for the view that branch production is governed by some kind of control mechanism (Katz *et al*., 1972). The branching patterns of trees and streams have been analysed by Leopold (1971); the position and size of branches were defined and a logarithmic relationship was found between branch order and the number and average lengths of the branches. This work has been extended and the methods applied to fungal branch systems. Gull (1975) investigated branching in *Thamnidium elegans*. Hyphal branching systems in small colonies of the fungus were analysed and each hyphal branch was given an order number. First order hyphae bear no branches. Second order hyphae bear only first order branches; each second order hypha is considered to extend to the tip of its longest component branch. A third order hypha bears first and second order branches. The results of the analysis are shown in Fig. 21 and support the view that there is a definite pattern of branching in many fungi. Very little is understood at present concerning the control of branch pattern. There may be an internal control due to the existence of a maximum rate of hyphal-tip extension (Katz *et al*., 1972) or to the nuclear volume/cytoplasmic volume ratio. On the other hand, branch pattern may be influenced by the concentration gradients of particular metabolites in the immediate external environment of the colony.

An experiment in which one aspect of the developmental pattern of a colony can be modified is outlined in Flowsheet 16. This experiment introduces the concept of the polarity of lateral branch induction normally observed in septate fungi. In *Geotrichum candidum* lateral branches are usually produced immediately behind a septum, i.e. at the distal end of each cell forming a branch. Very rarely will a branch be produced in any other position. Each branch grows towards the colony margin. It will have been noted (Section 2.7) that when growth of *G. candidum* was limited by oxygen the hyphae rarely produced lateral branches in the region beneath the centre of the coverslip. This is the starting point of the experiment in Flowsheet 16. When the coverslip is removed, and oxygen ceases to limit growth, lateral branches will form from these elongated cells. There is a breakdown in polarity as branches formed by some of the cells arise from the proximal end of the cells and grow towards the centre of the colony (Fig. 22). On the other hand, branches formed from the more proximal cells are often produced in the usual position (the distal end of the cell). How can this breakdown in pattern be explained? It could be argued that this disruption of the normal pattern of branch production is due to a higher than usual concentration of nutrients being maintained behind the colony margin due to the mycelial growth having been oxygen-limited. Is this high nutrient concentration inducing a reversal of the normal polarity? Further possibilities exist. The reversal in polarity may be due to the relatively higher metabolic activity of the peripheral region of the colony formerly beneath the coverslip. There is a higher density of mycelium here as the marginal hyphae approach the edge of the coverslip and lateral branch formation ensues in response to the higher oxygen tension. This could result in an enhanced production of various metabolites including one or more which may influence both the point of origin of the lateral branch and its subsequent direction of growth. There is

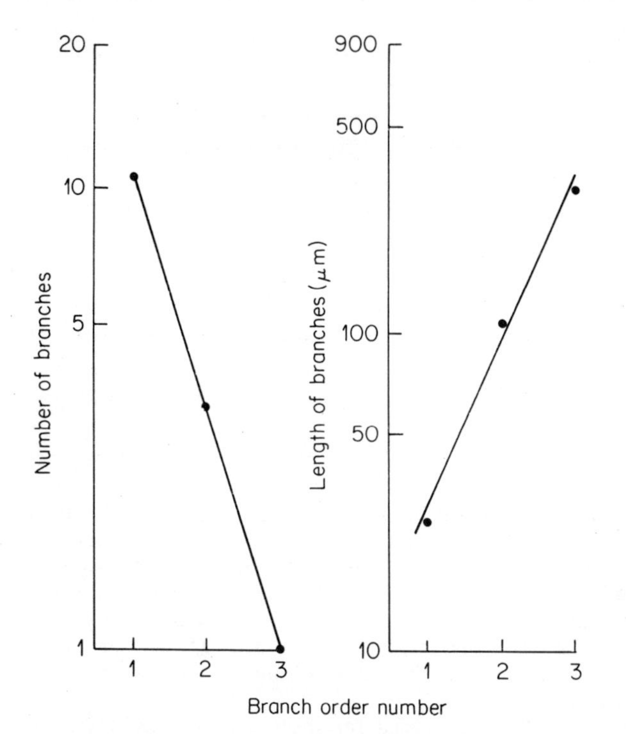

Fig. 21. Characteristics of a third order mycelial system of *Thamnidium elegans.* The numbers and average lengths of hyphal branches of the various orders are plotted. There is a logarithmic relationship between branch order and (a) the number of, and (b) the length of branches. The graphs produce two parameters which are useful in the analysis of a mycelium. The first is 'branch ratio'; this is the ratio of the number of first order branches to each second order one. The branch ratio is 3.8 and indicates that each branch of a given order will have approximately 3.8 branches of the next lower order. The second parameter is 'length ratio' which has a value of 4.0, i.e. in this mycelium any branch will tend to be four times as long as the average length of its branches of the next lower order. Reproduced, with permission, from Gull, *Transactions of the British Mycological Society,* **64**, 321–324 (1975). Published by Cambridge University Press.

another possibility. During normal development of hyphae the most proximal cells are usually the first to form lateral branches. Branching proceeds sequentially and the distal cells are the last to branch. This sequence would be expected as new cells are continually laid down by the advancing apical cell. The production of the lateral branch from a distal position could be due to the distal region of the cell being younger than the proximal region. Although the age difference is only likely to be of the order of one hour it is known that branch production in some fungi is

Flowsheet 16. Reversal of polarity of lateral branch formation in *Geotrichum candidum*

Inoculate plates of 1.0 per cent malt agar with arthrospores of *G. candidum*. Inoculate as a diametric streak. Incubate for 3 days at 25 °C. Place a coverslip on the margin of each colony as described in Flowsheet 14. Incubate some plates at 25 °C for a further day and some for a further 2 days. Remove the coverslip (carefully), incubate the culture at 30 °C and note the behaviour (at intervals of 0.5 h) of the area of the colony which was formerly covered. In particular, study the lateral branches which form. Is there any difference between cultures incubated for (a) one additional day, and (b) two additional days after removal of the coverslip? Is there any difference between hyphae at different depths in the growth medium?

Notes:
1. While not directly relevant to this exercise it is of interest to study the behaviour of the marginal hyphae which ceased extension-growth when the coverslip was applied. As stated in Section 2.7, these hyphae become vacuolate, the vacuoles extending to the hyphal apex. When the coverslip is removed the vacuoles regress and extension-growth may resume for a short time in some of the hyphae. After regression of the vacuoles, septa form in the apical cell of each hypha and arthrospores are delimited. This observation is relevant to Section 3.5 (Phase transformations).
2. If plastic Petri dishes are used for these experiments it may be noticed that hyphal cells beneath the central region of the coverslip do in fact form lateral branches when they grow at the bottom of the medium. This is because the plastic dish is permeable to oxygen. Although the amount of oxygen diffusing through the base of the dish is probably small it is nevertheless sufficient to allow some of the cells to branch. This effect is confined to the hyphae immediately adjacent to the base of the dish and this region should be discounted. While this problem can be completely avoided by using glass Petri dishes it is not considered of sufficient magnitude to warrant this change. Note the tropic responses of any lateral branches which form beneath the central region of the coverslip and at the bottom of the media. How do these branches differ from branches formed under 'normal conditions', i.e. at the colony surface in a region not covered with a coverslip?

favoured in cell regions which are relatively young. By preventing lateral branch formation for one or more days this polarity in age may be reduced within the cell so reducing the likelihood of a branch at the distal end. All these explanations are suggested to illustrate just a few of the numerous possibilities.

This exercise demonstrates one technique by which the normal pattern of colony morphology can be easily modified. The interpretation is far from simple.

3.4 CRYPTIC GROWTH

When a fungus is grown in a limited amount of culture medium growth stops in the sense that the dry weight of the culture ceases to increase and may even decline gradually. There is not an overall cessation of growth as some hyphae continue to grow in spite of the decline in dry weight of the culture. Also, some of the spores

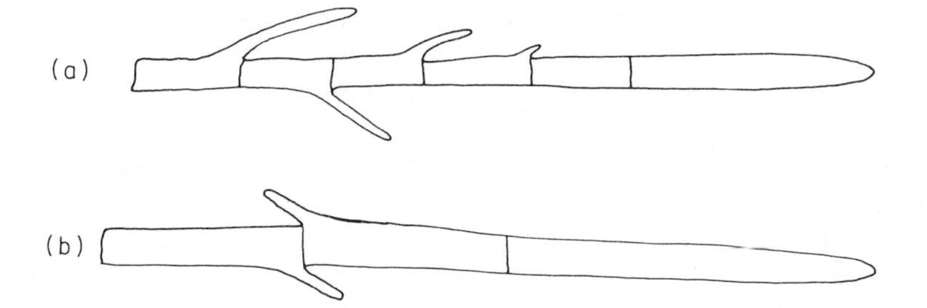

Fig. 22. Disruption of polarity of lateral branch induction in *Geotrichum candidum* on 1 per cent malt agar. (a) Normal polarity, (b) example of change in polarity observed after transfer of hyphae from oxygen-limiting to non-oxygen-limiting growth conditions. In (a) each branch is formed from the distal end of a cell. The branches nearest the hyphal tip are shorter as they are the most recently formed. In (b) the cells are longer and broader, and the branches are similar in length as they form simultaneously when the oxygen-limiting conditions are removed. Branch formation occurs from either end of each cell. Reproduced, with permission, from Robinson and Griffith, *Transactions of the British Mycological Society,* **68**, 311–314 (1977). Published by Cambridge University Press.

present may germinate. Ageing of the culture is a dynamic process and this can be demonstrated in the experiments outlined in Flowsheet 17.

In the first experiment a detailed examination of an old plate culture of *Fusarium oxysporum* is made over a period of several days (Park, 1961). Drawings are made of a small area of the culture and any changes noted. Wherever possible, particular configurations of spores can act as markers. Various changes may be observed such as the growth and disappearance of hyphae, the production of microconidia and chlamydospores, and the subsequent germination of some of them (Fig. 23).

The second experiment illustrates cryptic growth in *Geotrichum candidum* under different culture conditions from those of Experiment (1). A self-inhibitory arthrospore suspension in distilled water is studied over a period of several weeks. After a while a few arthrospores germinate and form minute colonies which in turn produce arthrospores. After several weeks most of the original arthrospores are not viable but recently-produced, viable arthrospores are present as a result of cryptic growth. A situation prevails in which arthrospores of the original suspension die and so reduce the effective arthrospore density such that some arthrospores are able to germinate. A certain amount of nutrient material is released by the non-viable and autolysing arthrospores and this, as well as the resultant decline in arthrospore density, helps trigger germination in the remainder of the population. The result is a continuous activity in the arthrospore population with arthrospores dying and releasing nutrients, and arthrospores germinating and producing more arthrospores. At any one time there will be a mixture of arthrospores of various ages but, with time, the number of viable arthrospores in the population declines.

In Experiment (3) arthrospores are seeded at a low density in distilled water agar

Flowsheet 17. Cryptic growth in *Fusarium oxysporum* and *Geotrichum candidum*

Experiment (1) Inoculate a plate of S agar medium with *F. oxysporum*. Incubate for 2 weeks at 25 °C and then note changes of selected staled areas of each plate at selected time intervals (at least 1 day). Pour deep plates for this experiment as the culture will lose a lot of moisture over a 2-week period.

Experiment (2) Prepare a spore suspension of *G. candidum* by adding 10 ml sterile distilled water to a 3-day-old slope culture on malt agar. Shake gently for 10 s and decant the suspension into a sterile McCartney bottle. Adjust the suspension as necessary with sterile distilled water to give 10^6 spores ml^{-1}. Transfer 2 ml aliquots to sterile McCartney bottles, screw down the caps tightly and seal with molten paraffin wax. Incubate at 25 °C and examine at weekly intervals for 6 weeks, commencing 2 weeks after setting up the experiment. Examine by pipetting a 0.1 ml sample from a bottle to a plate of 1 per cent malt agar. Spread the sample over the plate and incubate at 25 °C; make observations after 4 h.

Experiment (3) Prepare a spore suspension of *G. candidum* as in Experiment (2) and dilute to give a density of 10^4 spores ml^{-1}. Add 1 ml aliquots of suspension to 9 ml aliquots of sterile, molten (50 °C) distilled water agar in McCartney bottles. Mix and pour the contents of each bottle into a Petri dish. Allow a small amount of molten paraffin wax to run around the perimeter of the inside of each lid and replace the inverted base of each dish while the wax is semi-solid. Press the two halves of each dish gently together to make an air-tight seal. Incubate plates upside down at 25 °C for at least 7 days to allow the spores to germinate and the small colonies to form spores. Examine and draw individual colonies at weekly intervals for 6 weeks, commencing 1 week after setting up the experiment. Do not open the dishes. The colonies can be examined (x10 objective) by placing the plate (inverted, as for incubation) on the microscope stage.

Notes:
It is important for Experiments (1) and (3) to be able to recognize particular areas of a culture over a long period of time. It is convenient to etch small circles on the bases of the dishes in Experiment (3) and to identify these with letters or numbers. The individual colonies within each area can be sketched and any changes noted. In Experiment (1) small particles of sterile, coloured glass (or other inert material) can be dropped onto the colony surface before observation to act as initial markers.

and allowed to germinate and form small colonies. This illustrates another facet of cryptic growth. Each hypha of each colony ultimately ceases growth and forms arthrospores and a static situation appears to prevail for some time. However, the arthrospores persist for longer than the hyphae to which they are attached. The hyphal contents lyse but the cell walls remain. At a later stage it is common for one or more arthrospores in the chains to germinate and produce hyphae which ultimately cease growth and produce more arthrospores. The situation is not as static as might be imagined.

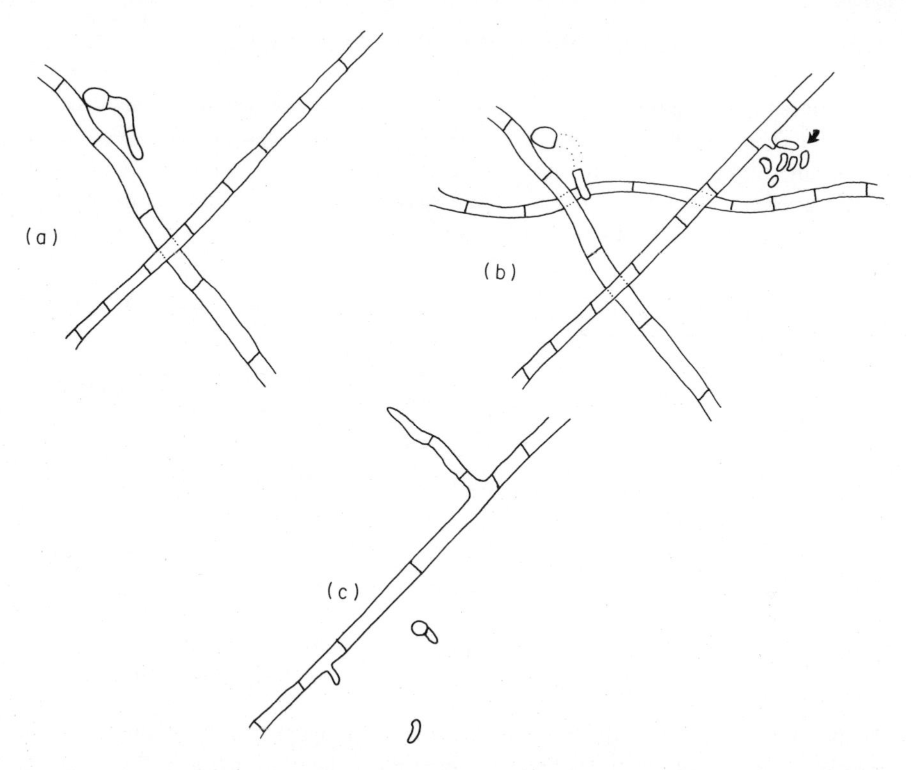

Fig. 23. Cryptic growth in *Fusarium oxysporum* on S agar. (a) Drawing of an area of a staled agar culture. (b) The same area 22 h later, and (c) after a further 30 h. Note the newly-formed hypha in (b) and the formation of a clump of microconidia (arrowed). In (c) many of the structures present in (b) have autolysed. Reproduced, with permission, from Park, *Transactions of the British Mycological Society*, **44**, 377–390 (1961). Published by Cambridge University Press

3.5 PHASE TRANSFORMATIONS

Fungal hyphae may become modified and form spores. The transition from the somatic phase to the reproductive phase and vice versa is termed 'phase transformation' and can be readily studied in colonies of many spore-bearing fungi. The marginal hyphae do not usually sporulate and spores are normally produced by lateral hyphae growing behind the margin. The distance from the margin at which spores form is quite precise for any given species under defined growth conditions on solid media. This process will have been observed in Section 3.1 using *Fusarium oxysporum* as the test fungus.

Spore induction can be studied readily in *Geotrichum candidum* using an extension of the self-inhibition experiments described in Section 1.3. Details are given in Flowsheet 18. In Experiment (1) arthrospores seeded at different densities in distilled water agar are left to develop. An interesting effect emerges. The higher the sowing density the lower the percentage germination of the arthrospore population; the germinating arthrospores themselves form arthrospores after a very

Flowsheet 18. Phase transformations in *Geotrichum candidum*

Experiment (1) Prepare a spore suspension of *G. candidum* in sterile distilled water. The suspension should have a density of 5×10^6 spores ml^{-1}. This can be obtained by adding 10 ml sterile distilled water to a 3-day-old malt agar slope of *G. candidum* (incubated at 25 °C) and shaking for 10 s. Decant the suspension, adjust the density as necessary using a haemocytometer, and proceed as follows:

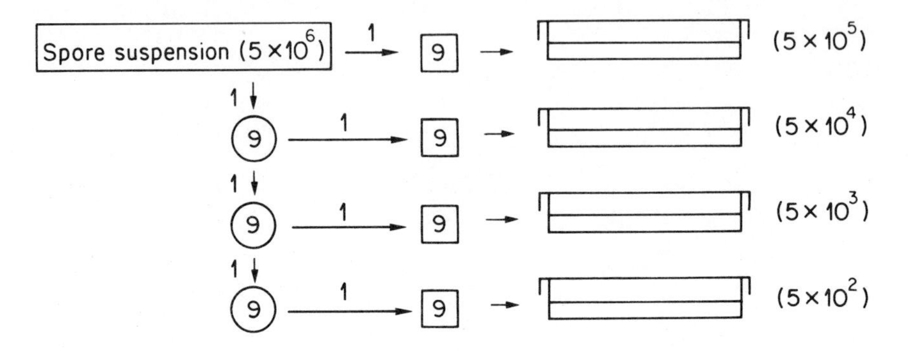

Legend: ○ = no. ml sterile distilled water in McCartney bottle
□ | = no. ml sterile distilled water agar in McCartney bottle

Numbers in brackets relate to spore densities (no. spores ml^{-1}); numbers not in brackets refer to volume (ml) of spore suspension to be transferred.

Notes:
1. Melt the McCartney bottles of distilled water agar in an autoclave and then store in a water bath at 50 °C prior to use.
2. Incubate all plates at 25 °C and examine when growth has ceased, i.e. when all the hyphae have formed arthrospores. This will take 3–4 days at the lower sowing densities.
3. Determine how (a) number of arthrospores/hypha, and (b) number of hyphae/colony vary with the sowing density.

Experiment (2) Inoculate a plate of S agar centrally with *G. candidum*. Inoculate a second plate similarly but on the surface of a 5 cm square of Cellophane placed on the agar surface. See Flowsheet 1 for details concerning preparation of the Cellophane. Incubate plates for 4 days at 25 °C.

↓

Remove the Cellophane square carefully by lifting one corner with sterile forceps. Place the Cellophane and its attached colony over the colony of *G. candidum* on the other plate so that a portion of the colony margins overlap. The overlapping region should be approximately 1 cm on a line joining the centres of the 2 colonies.

↓

Observe the events in the overlapping region after 3 and 24 h.

short period of growth. At lower sowing densities almost all the arthrospores germinate and the young colonies develop more extensively before forming arthrospores. More branches and more arthrospores per branch are formed in colonies developed from arthrospores sown at low densities. Every hypha ultimately forms arthrospores. As well as demonstrating sporulation this exercise also demonstrates self-inhibition between colonies of fungi. Very rarely do the hyphae of neighbouring colonies overlap.

In Experiment (2) a colony of *G. candidum* is placed so as to overlap a second colony of the same species. After a few hours it will be seen that the leader hyphae which overlap cease growth and sporulate. Experiments of this type have been used as evidence for the production by fungi of metabolites which can induce morphogenetic effects similar to those which occur naturally in hyphae in the more central regions of the colony. It could be argued equally that the sporulation observed in this experiment is due primarily to increased competition for, and the resulting depletion of, available nutrients in the overlapping region. This would result in a lowering of the specific growth rate of the marginal hyphae and sporulation would be favoured.

4
Reproduction in Fungi

Most species of fungi are able to reproduce asexually and sexually. Several methods of asexual reproduction are found in fungi. These include fragmentation of a thallus, each fragment growing into a new colony; budding of somatic cells or spores, each bud producing a new individual; fission of somatic cells into daughter cells; and spore production, each spore having the potential to germinate and develop into a new mycelium. Fragmentation of hyphae is a normal means of propagation in some fungi and the fragments are called arthrospores. *Geotrichum candidum*, the subject of many experiments in this book, forms arthrospores. In the laboratory many fungal cultures are kept growing by transferring pieces of mycelia to fresh media, this is also an example of mycelial fragmentation. Budding and fission are characteristic of many yeasts. Budding is the formation of a small outgrowth (bud) from a parent cell. As the bud forms, the nucleus of the parent cell divides and one daughter nucleus enters the bud. The bud increases in size and eventually breaks off and forms a new individual. The buds do not always break off and in some species a chain of buds is produced. Fission is the splitting of a cell into two daughter cells; a constriction appears in the cell when it is about to divide and a new cell wall is formed along the constriction to separate the original cell into the two component daughter cells.

Spore production is the most common means of asexual reproduction in the fungi. Spores from different species of fungi vary in colour, in size, in number of cells, and in the arrangement of the cells. Spores also vary in the way they are borne. Asexual spores are either produced in sporangia and are called sporangiospores, or are produced at the tips or sides of hyphae and are called conidia. Sporangiospores may be motile or non-motile. In the lower fungi the sporangiospores are usually motile and are called zoospores. When non-motile they are called aplanospores.

Sexual reproduction involves the union of two compatible nuclei. There are normally three stages in sexual reproduction: plasmogamy, karyogamy, and meiosis. In plasmogamy two protoplasts unite to bring the compatible nuclei closer together within the same cell. Karyogamy is the fusion of the two nuclei. In the lower fungi karyogamy occurs almost immediately after plasmogamy but in higher fungi these two processes may be separated in time and space, plasmogamy resulting in a binucleate cell. The two nuclei (a dikaryon) may not fuse until much later and can be perpetuated from cell to cell by the synchronous division of the two associated nuclei. When nuclear fusion occurs it is followed by meiosis which constitutes the third phase of sexual reproduction and restores the haploid condition in the four nuclei which result.

Before studying reproduction in fungi it is important to gain experience in the preparation of temporary mounts of hyphae and spores from growing cultures. Some ideas are set out in Flowsheet 19. The techniques employed have been selected because of their relevance to the later practical work suggested in this chapter. A small amount of the practical work suggested in this chapter is not concerned directly with reproduction but is included out of general interest.

Flowsheet 19. Examination of fungal morphology and preparation of temporary mounts

---

(1) *Germinating spores.* Spread sporangiospores of *Mucor plumbeus* and conidia of *Botrytis cinerea* on separate plates of 1 per cent malt agar. Incubate the plates for approximately 18 h at 25 °C and cut 1 cm squares from the medium with a scalpel or razor blade. Mount each square on a microscope slide. Draw various stages in the germination process and pay particular attention to the size of the spores (germinated and non-germinated), the presence of cross-walls and vacuoles. Use the x 10 objective.

(2) *Young colonies.* Inoculate separate plates of 1 per cent malt agar with spores of *M. plumbeus* and *B. cinerea.* Incubate for 30 h at 25 °C by which time young colonies will have formed. Examine with the x 10 and x 20 objective. How do the sizes of hyphae (width and, where appropriate, compartment length) compare with those in (1)? Note patterns of branching and septation. Make annotated drawings of your observations.

(3) *Spores and sporulation.* Inoculate separate plates of 1 per cent malt agar with spores of *Trichothecium roseum* and *M. plumbeus.* Examine colonies of each fungus after incubation at 25 °C for 14 days. Inoculate the centre of each plate so that a single colony develops. Examine as follows:
   (a) Put a whole plate (less lid) on the stage of the microscope and examine the culture by transmitted and by reflected light. Use a x 10 objective. Alternatively, a portion cut from the culture and placed on a microscope slide can be used in this dry mount method. A x 40 objective can yield additional useful information but only if used with considerable care; it is important to avoid letting the front lens touch either the agar or the fungus.
   (b) Take some of the aerial growth of the fungus on an inoculating needle and place it in a drop of water on a microscope slide. Tease out the mycelium, add a coverslip and examine.
   (c) As (b) but mount in lactophenol.
   (d) As (b) but mount in lactophenol + acid fuchsin.
   (e) Cut out a small portion of agar plus fungus using a flamed scalpel. Mount in lactophenol + acid fuchsin, add a coverslip and examine.

(f) As in (e) but mount in ammoniacal congo red. Examine for cell walls and septa.

(g) Cut a length (5–6 cm) of Sellotape. Using gentle finger pressure touch the central 1 cm of the strip (sticky side downwards) on the colony to be examined. Place the strip (sticky side down) on a microscope slide with the colony impression over a drop of lactophenol + acid fuchsin. Examine for details of sporulating structures.

In this series of examinations note:

(i) Some fungal structures are not easily wetted, and tend to trap air.

(ii) Sporulating structures are often very delicate and may not survive the handling involved in making a wet mount.

(iii) Fungal material is often of low contrast and stain-incorporated mountants may be of considerable benefit.

During your examinations attempt to determine how the spores are borne on the sporophore. Also, compare the size of ungerminated spores of *M. plumbeus* with the germinated spores in (1). What could account for any differences that you observe?

Notes:

1. *Permanent mounts.* While lactophenol mounts are so easily made as temporary preparations they can with a little more effort be made into satisfactory permanent mounts. To do this the edge of the coverslip is sealed to the slide with a cement which is not attacked by the corrosive lactophenol. Ordinary (colourless) finger nail varnish is suitable and Gurr's Glyceel (Searle Diagnostic, High Wycombe, Bucks., England) may be used. Points to observe are that the slide and upper surface of the coverslip are clean and not greasy, that the amount of lactophenol used is small and does not spread out from below the coverslip, and that the painted-on strip of sealant is wide enough (2–3 mm). Allow the sealant to set with the slide in a horizontal position.

2. *Reagents*

Lactophenol: phenol, 10 g; lactic acid, 10 g; glycerol, 20 g; distilled water, 10 ml. Mix the phenol with the water until dissolved and then add the lactic acid and glycerol.

Lactophenol + acid fuchsin: add acid fuchsin, 0.5 g, to the formula for lactophenol.

Ammoniacal congo red: congo red, 1 g; ammonium hydroxide (specific gravity; 0.88), 5 drops; distilled water, 100 ml.

4.1 PHYCOMYCETES

The sporangia-bearing fungi or Phycomycetes have in common the lack of a regular system of perforate septa to divide the mycelium into compartments. Complete septa may develop on wounding, with age, or in connection with delimitation of reproductive structures. Despite this common anatomical feature there are several very distinct and evolutionarily diverse groups included here. Two main classes of fungi, the Oomycetes and Zygomycetes, illustrate the Phycomycetes. Fungi of the

Oomycetes have a diploid mycelium with the cell walls containing a fair proportion of cellulose. Asexual reproduction is by motile biflagellate zoospores and sexual reproduction is by production of oospores. Fungi in the class Zygomycetes have a haploid mycelium with cell walls lacking cellulose; chitin is the characteristic cell-wall carbohydrate here. Asexual reproduction is by non-motile spores (sporangiospores), sexual reproduction is by zygospores.

Oomycetes

Asexual reproduction

Colonies of almost any species of *Saprolegnia* are suitable for the study of asexual reproduction in the Oomycetes. Colonies of the fungus can be grown either on solid media or on baits of hemp seed suspended in liquid media. When grown on a bait the fungus produces fine feeding hyphae which invade the substrate and

Flowsheet 20. Asexual reproduction in *Saprolegnia*

Pour approximately 10 ml molten 1 per cent malt agar into a Petri dish and inoculate centrally with a species of *Saprolegnia*, e.g. *S. ferax*. Incubate the culture at 20 °C or 25 °C for at least 2 days.

↓

Cut discs (5 mm diameter) from a portion of the colony at least 1 cm inside the colony margin. Use a flamed cork borer to cut the discs.

↓

Transfer a single disc to a sterile 9 cm diameter Petri dish and add 15 ml sterile distilled water. Incubate for 18 h at laboratory temperature (20 °C).

↓

Examine the culture with a microscope using a x 10 objective. Make notes on the various stages of sporangial formation observed. Note the presence of discharged sporangia and look for sporangia which are discharging zoospores. Observe the motile zoospores; some of the zoospores may have settled on the base of the Petri dish and produced germ-tubes.

Notes:

1. Species of *Achlya* can be substituted for *Saprolegnia.* If zoospore production is very low the number of zoospores can sometimes be increased by flooding the culture discs with sterile, filter-paper-filtered river water instead of distilled water. Zoospore production can often be stimulated by the replenishment of the flooding solution during the incubation period.
2. It is recommended that discs are not cut at the colony margin as on occasions this has resulted in low yields of zoospores.
3. Observation is facilitated by pouring off some of the flooding solution prior to observation.

which are not readily seen. A system of very broad hyphae is soon produced and this radiates from the substrate into the water. During the first few days these broad hyphae produce elongate sporangia in which zoospores differentiate. The zoospores escape at maturity through a terminal opening and swim away. They are capable of germinating to colonize fresh substrates. Sporangial production and discharge are promoted by a change of water in the culture. A convenient technique for studying asexual reproduction is outlined in Flowsheet 20.

Sexual reproduction

Sexual reproduction in the Oomycetes can be demonstrated readily in *Achlya ambisexualis.* Inoculate 1 per cent malt agar plates with both the male and female strains of *A. ambisexualis.* The points of inoculation should be about 4 cm apart. When the two colonies meet sexual reproduction occurs and different stages in the sexual process should be present. Look for stages in the development of oogonial branches, antheridial branches, oogonial initials, oogonia, antheridia, oospheres, and oospores. Correlate your sequence with that described by Raper (1952) in connection with his hypothesis of hormonal control. A brief summary of some of Raper's experiments follows.

Raper (1940) demonstrated that most of the sexual events in *Achlya* are based on hormonal secretions. He placed male and female mycelia in liquid culture in separate small cells and drew a stream of water sequentially through the cells via small connecting siphons. In the first cell a female mycelium was placed, in the second a male mycelium, in the third a female, and in the fourth a male. Water was siphoned from cell one to cell two and so on. The male mycelium in the second cell formed antheridial branches and, similarly, the male mycelium in the fourth cell reacted some hours later. The female mycelium in the third cell produced oogonial initials which did not develop further. The antheridial hyphae in the fourth cell grew towards the tip of the siphon from the third cell. The antheridial hyphae in the second cell did not grow towards the siphon carrying water from the first cell, nor did oogonia develop on the female mycelium in the first cell. Raper postulated four hormones (A, B, C, and D) to explain these results but subsequent work by Barksdale (1963) challenged the existence of as many as four hormones. Two hormones have now been characterized and named antheridiol and oogoniol. A simplified scheme of sexual reproduction in *Achlya* is presented in Fig. 24 and is based on the hormones which have been characterized.

To return to the plate culture technique suggested for the study of these sexual interactions. When a male and female mycelium approach each other the first visible response is the formation of branched antheridial hyphae on the male. These branches are thinner than the vegetative hyphae and grow towards oogonial initials which develop as spherical organs on the female mycelium. The antheridial hyphae curve around the oogonial initials and cross-walls form near the tips of the antheridial hyphae. The delimited cells are the antheridia. A cross-wall forms at the

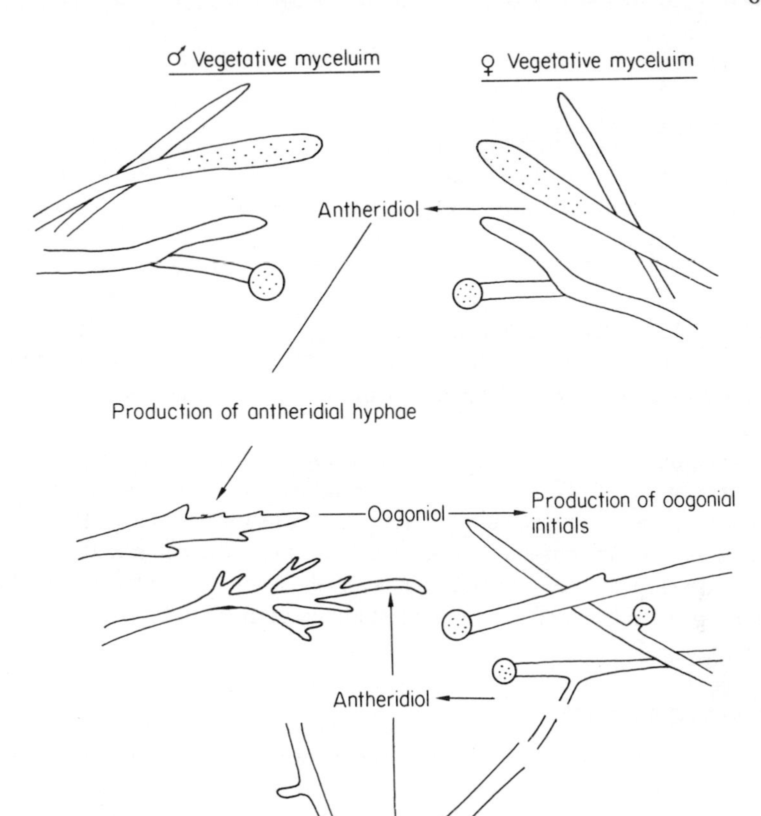

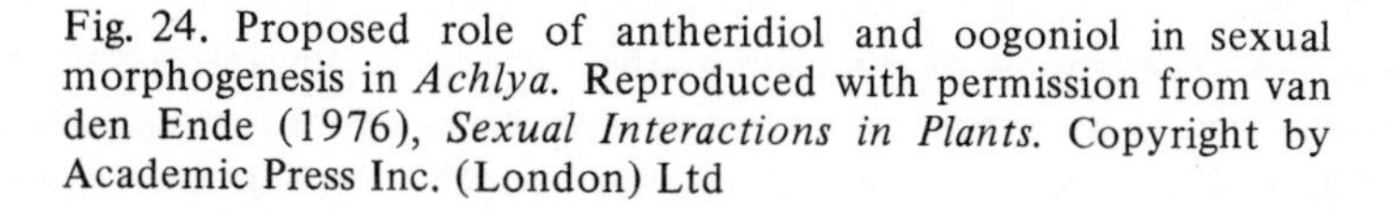

Fig. 24. Proposed role of antheridiol and oogoniol in sexual morphogenesis in *Achlya.* Reproduced with permission from van den Ende (1976), *Sexual Interactions in Plants.* Copyright by Academic Press Inc. (London) Ltd

base of each oogonial initial and within each oogonium the protoplasm reorganizes to form one to twenty mononucleate eggs. The antheridium penetrates the oogonial wall with mononucleate fertilization tubes, each of which fuses with one egg. Meiosis probably occurs in the sex organs and not in the fertilized egg, the vegetative mycelia being diploid (Barksdale, 1966). The most striking aspect of this sexual morphogenesis is the sequence of events which suggests a successive secretion of substances which control the morphological changes (Raper, 1952).

Zygomycetes

Asexual reproduction

Inoculate a plate of 1 per cent malt agar with sporangiospores of *Mucor plumbeus.* Incubate for 7 days at 25 °C. Examine the culture as follows, paying particular attention to the asexual reproductive apparatus. Remove the lid of the Petri dish and observe with the x 10 objective before the culture begins to dry out. Observe the size of the sporangia and look for the very fine needle-shaped crystals which encrust each sporangium. Mount a portion of the culture in lactophenol and examine in detail noting the columella and collarette. Examine young developing sporangia as well. What happens to the mature sporangial wall when mounted in liquid?

Examine 4- to 5-day-old cultures of *Rhizopus stolonifer* grown on 1 per cent malt agar at 25 °C. Use the same technique as for *M. plumbeus.* Compare the reproductive structures and their behaviour in relation to water. Allow a plate of *R. stolonifer* to dry out and notice what happens to the columellae. Observe stolons and development of rhizoids. Some of the spore characteristics in *Mucor* and *Rhizopus* can be compared by the techniques outlined in Flowsheet 21.

Sexual reproduction

Blakeslee (1904) first described sexual reproduction in the Mucorales and his findings can be confirmed as follows. Grow strains of each mating type, i.e. *plus* and *minus* strains, of *Mucor mucedo* on (a) separate plates, and (b) the same plate of 1 per cent malt agar for 7 days at 25 °C. If the cultures from (a) are examined it will be seen that they appear identical in morphology. Both strains form aerial hyphae which are narrow and easily distinguished from the stout hyphae which are destined to become sporangiophores. When both strains are inoculated 4 cm apart on the same plate (b) a series of changes occurs as the mycelia of the opposite strains grow towards each other. Hyphae are produced which are intermediate in size between vegetative hyphae and sporangiophores. These hyphae are called zygophores. Zygophores of opposite mating type grow towards each other by a process of mutual attraction. When two zygophores of opposite strains contact each other a swollen outgrowth (the progametangium) develops on each zygophore at the point of contact. The *plus* and *minus* progametangia adhere to each other and, when they have enlarged considerably, septa form which cut off terminal cells (the gametangia). The gametangia fuse and form the zygote which rounds off and forms a dark, denticulated wall. The mature fusion cell is called the zygospore.

Banbury (1954) and Plempel (1963) have shown that *plus* and *minus* mycelia of *M. mucedo* grown together in liquid culture produce substances which induce zygophores in both strains of the fungus. The structural characterization of the hormones involved was accomplished by Caglioti *et al.* (1967) and Cainelli *et al.* (1967). These workers observed that a chloroform extract of submerged mated cultures of *Blakeslee trispora* contained acids which, when added to cultures of a

Flowsheet 21. Comparison of spore characteristics in *Mucor plumbeus* and *Rhizopus stolonifer*

Inoculate separate plates of 1 per cent malt agar with *M. plumbeus* and *R. stolonifer*. Incubate for 7 days at 25 °C.

Experiment (1) Take a fresh plate culture of each fungus. Wet a hair with ethanol and then with water, next drag the hair across one of the cultures. Mount the hair in a drop of lactophenol on a slide. Repeat this procedure for the other species. Assess the spore harvest under these moist conditions.

↓

Repeat the procedure above but this time use cultures which have been dried by exposure to the laboratory atmosphere for about 0.5 h and use a dry hair. Compare this dry spore harvest with that under moist conditions.

Experiment (2) Breathe gently on a drying plate of *R. stolonifer* and, at the same time, observe the culture under a x 10 objective or a binocular dissecting microscope. Look for hygroscopic movements connected with spore liberation.

Experiment (3) Add water (approximately 5 ml) to a plate of each species. Decant the water from each plate into a separate test-tube and examine for spores. Note the partition of the spores between the main bulk of the liquid and the surface film (meniscus). What can you conclude about the relative degrees of wettability of the two species?

---

single strain of this fungus, increased the yield of carotenoids. Van den Ende (1968) showed that these substances, called trisporic acids, were identical to the zygophore-inducing substances of this species and of *M. mucedo*. The trisporic acids induce zygophore formation in both mating types of *M. mucedo* and this lack of mating-type specificity is not paralleled by any other plant sex hormone so far studied. The zygophores of opposite mating types grow towards one another and Mesland *et al.* (1974) noted that zygophores are not only attracted by zygophores of the other mating type but also by vegetative mycelium and sporangiophores of the opposite strain. The active substances have not been characterized.

4.2 ASCOMYCETES

This class of fungi is characterized by having its sexual process culminating in the formation of ascospores which are produced in a definite number (usually eight) in an ascus. The ascus is a single cell, in fact a special type of sporangium. In one sub-class, the Hemiascomycetes, asci are produced singly (not grouped) and are not enclosed by sterile hyphae. Some members of this sub-class will be treated later, as yeasts. The remainder of the Ascomycetes, the Euascomycetes, have the asci aggregated in and protected by an ascocarp.

Typically, the female cell, the ascogonium, contains initially a single nucleus. In one of several ways a single male nucleus is introduced into the ascogonium. The phenomenon where a cell contains two compatible haploid nuclei is dikaryosis. The process by which it is brought about is dikaryotization (not fertilization: there has been no nuclear fusion at this stage). In these fungi, nuclear fusion, or karyogamy, is separated in time from cytoplasmic fusion, or plasmogamy. The separating interval is bridged by the nuclear phase known as the dikaryon.

The dikaryotic ascogonium produces a number of branches, each cell of which possesses two nuclei. The branches are the dikaryotic ascogenous hyphae. In the most highly organized ascocarps they form a basal mat that eventually produces the hymenial (fertile) layer. The rest of the ascocarp is formed from the growth of sterile enveloping hyphae produced by the haploid cells near the dikaryotized ascogonial cell. These sterile hyphae aggregate to form the firm, sometimes hard and compact, ascocarp which is a structure of definite form.

By a process known as 'crozier formation' the penultimate cell of each ascogenous hypha can become an ascus (Fig. 25). In it the two compatible nuclei fuse to become a single diploid nucleus. There are no mitotic divisions of this diploid phase. It exists as a single nuclear generation and its next development is to undergo meiosis. Each of the four resultant haploid products of meiosis may then divide by a normal mitosis so that eight nuclei are present.

Ascospore formation occurs by a phenomenon known as 'free cell formation' which differs from a progressive cleavage of the cytoplasm about the nuclei to form

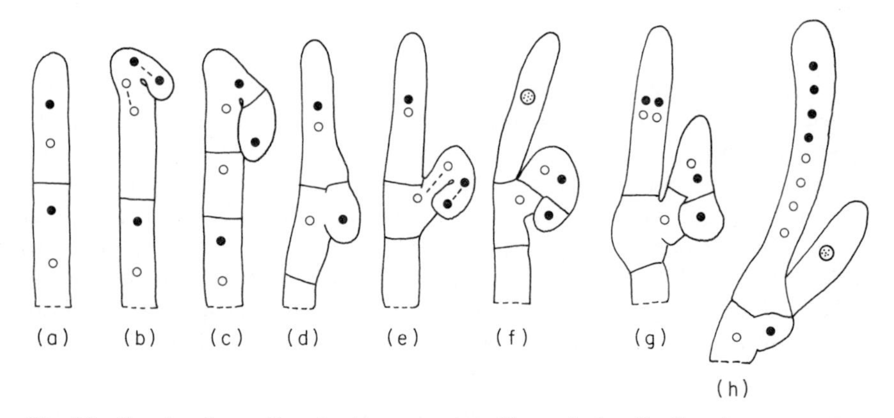

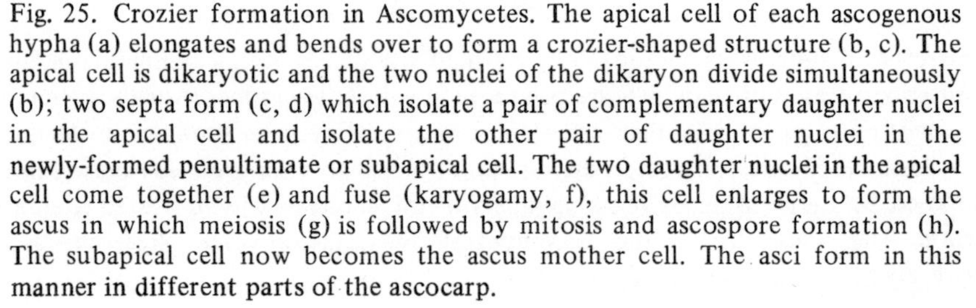

Fig. 25. Crozier formation in Ascomycetes. The apical cell of each ascogenous hypha (a) elongates and bends over to form a crozier-shaped structure (b, c). The apical cell is dikaryotic and the two nuclei of the dikaryon divide simultaneously (b); two septa form (c, d) which isolate a pair of complementary daughter nuclei in the apical cell and isolate the other pair of daughter nuclei in the newly-formed penultimate or subapical cell. The two daughter nuclei in the apical cell come together (e) and fuse (karyogamy, f), this cell enlarges to form the ascus in which meiosis (g) is followed by mitosis and ascospore formation (h). The subapical cell now becomes the ascus mother cell. The asci form in this manner in different parts of the ascocarp.

cells (as happens in most sporangia) in that not all the cytoplasm is included in the resulting spores. Cell walls form simultaneously about the nuclei and include some of the cytoplasm within the (usually uninucleate) spores, but some is excluded. The portion left outside may contribute to the characteristic sculpturing of the outer wall of the ascospores, but probably its main function is to contribute material which will provide energy for the subsequent active liberation of the ascospores from the ascocarp.

Euascomycetes

Sordaria fimicola. Inoculate plates of 1 per cent malt agar centrally with *S. fimicola.* Incubate the plates for 14 days at laboratory temperature in a situation where illumination is unilateral. The plate culture bears black perithecia, each with a short neck or beak which opens by an ostiole at the tip. Examine whole perithecia carefully, the beaks point towards the direction of the light. Have any spores been dispersed? Look on the lid of the dish. Mount several whole perithecia (if you are able to detach them whole without breaking the ascocarp wall at the base where it is attached to the substratum) in dilute iodine solution. Watch for ascospore discharge. How many spores are ejected at one time? Do they come out as a unit or one at a time? Make drawings of individual spores.

Mount a whole perithecium (or, preferably, several at different stages of development) in lactophenol + acid fuchsin, and cover with a coverslip. Press with a pencil butt or something similar on the coverslip immediately over a perithecium so as to burst it. Using the microscope, examine the perithecial contents. Draw stages in ascus development.

Neocosmospora vasinfecta. Inoculate plates of 1 per cent malt agar with *N. vasinfecta* and incubate for 14 days in the light at laboratory temperature. The construction of the perithecium in *N. vasinfecta* is very similar to that in *Sordaria.* Examine in the same way that you did for *Sordaria.* The mode of spore liberation is quite different. In *Sordaria* a mature ascus becomes turgid and elongates until its tip reaches the ostiole, whereupon the contents are shot out with some force, but in *N. vasinfecta* the walls of the mature asci autolyse liberating the spores and other material into the perithecial cavity. The non-spore material is hygroscopic and under humid conditions becomes a mucilaginous mass that swells and extrudes through the ostiole, carrying the ascospores with it. In dry conditions the mucilage hardens and prevents further movement of the spores; under moist conditions the spores may be washed downwards with water rivulets or be carried by splashed water droplets. They tend to adhere to surfaces.

Notice also, in the culture of *N. vasinfecta*, small hyaline, smooth, sausage-shaped conidia. (The ascospores are coloured, more-or-less spherical, and rough walled.) The conidia are the asexual spores. If an ascomycete were to produce conidia in this way, and could not be induced to produce ascocarps, it would be classified in the Fungi Imperfecti.

Chaetomium globosum. Inoculate plates of 1 per cent malt agar with *C. globosum* and incubate for 14 days in the light at laboratory temperature. *C. globosum* has perithecia which are surmounted by coiled or forked bristles. The perithecium may or may not have a beak and an ostiole. Despite some species lacking an ostiole the ascocarp is still considered a perithecium, not a cleistothecium. A cleistothecium is a closed ascocarp, but is also characterized by having the asci irregularly arranged within it; a perithecium has a definite hymenial layer – as in *C. globosum*.

The walls of the asci break down at maturity, as do those of *Neocosmospora*, but the liberation mechanism operates under dry conditions. If a perithecium is dragged along, as by a small animal whose body has tangled with the bristles, the spores are shaken out of the ostiole or out of the rupture in the base where the perithecium was torn away from the substratum.

Make drawings of a perithecium and of ascospores.

Aspergillus repens. A. repens is an osmophilic fungus and can be grown by inoculating autoclaved soaked prunes with conidia or ascospores. We are concerned here with the perfect, cleistothecial stage of the genus *Aspergillus. Aspergillus* is better known for its imperfect, conidial stages in the Fungi Imperfecti. The perfect stage was first designated *Eurotium* but for nomenclatorial reasons the genus is best kept together under the generic name *Aspergillus.*

Conidial areas are greenish in colour. Ignore these areas. Cleistothecia when mature are yellowish-orange. Pick off one or two cleistothecia and examine them as you did for ascocarps of the previous species. Note particularly the absence of a linear arrangement of spores in the ascus. Which of the other species you looked at showed this? What does it correlate with in terms of method of spore liberation? What is the three-dimensional shape of the ascospore of *Aspergillus*? In *A. janus* the outer, pseudoparenchymatous wall of the cleistothecium is built up from swollen and thickened hyphal tips (Hülle cells). With a fine needle take some mycelium from a part of the colony that shows a pale-yellow colour. Mount in lactophenol + acid fuchsin. Examine for ascogonial stages and early stages in cleistothecial development.

Hemiascomycetes and yeasts

Hemiascomycetes are those fungi which produce naked asci singly. The most significant order, the Endomycetales, has the zygote transformed directly into a single ascus (no dikaryotic phase and no ascogenous hyphae are produced). It includes both a mycelial and a unicellular series, with some very clear connections between the two. These fungi are the mycelial and non-mycelial ascosporogenous yeasts. Some yeasts, however, produce ascospores with difficulty; others are not known to produce them at all. Obviously a position for the latter yeasts in the Fungi Imperfecti is one possibility but where their relationships with perfect yeasts can be seen, as in morphological and biochemical features, they can be treated as imperfect or anascosporogenous yeasts. Again there is both a mycelial and a non-mycelial (unicellular) series.

Unicellular yeasts reproduce either by budding (a process whereby cytoplasm suddenly bursts through a hole in the parent cell wall and then secretes a cell wall about itself, later becoming separated by a cross-wall at the narrow neck or junction of the two cells) or by fission (in which growth is more gradual and general, and cell wall synthesis keeps phase with synthesis of cytoplasm). In either case the basically unicellular yeast may under unusual circumstances, e.g. in old, staled cultures or under antibiotic treatment, be induced to produce a mycelial or pseudomycelial colony form. Conversely, some of the mycelial yeasts may under some conditions grow as unicells (the phenomenon is known as dimorphism). Of these fungi, those which grow by budding when growing as unicells may, when growing in the mycelial phase, produce spores by a budding process (these bud-cells are called blastospores). Those mycelial yeasts which show fission when induced to grow as unicells normally sporulate from the mycelial phase by arthrospores.

Ascosporogenous yeasts

Eremascus albus. This mycelial yeast is osmophilic and fails to make any growth on the usual laboratory culture media. It can be grown on plates of 1 per cent malt agar + 40 per cent sucrose. Incubate cultures at 25 °C. Mount a portion of the colony in lactophenol and examine for and make drawings of copulation (in which two gametangial hyphae entwine) and for asci (eight ascospores). The fusion cell swells and forms the ascus which has four or eight ascospores.

Schizosaccharomyces octosporus. Inoculate 100 ml sterile 1 per cent malt broth (in a 250 ml flask) with cells of *S. octosporus.* Incubate at 25 °C for 2 to 3 days. This is a non-mycelial fission yeast which does not reproduce by budding. After some growth (elongation of the more or less cylindrical cell) of a young cell a cross-wall develops medianly across the long axis, cutting the cell into two. Separation of the daughter cells occurs by a split which develops along the mid-line of the cross-wall. Observe stages in growth by taking a sample from the broth culture and mounting directly under a coverslip.

S. octosporus forms ascospores with greater ease than do most yeasts. It is haploid in the somatic phase and fusions occur between pairs of cells. Four or eight ascospores then develop in the zygote. Put a drop of water on a clean slide. Touch the *centre* of a malt agar-grown colony with the tip of a flamed and cooled inoculating needle. Transfer the cells picked up to the drop of water and mix. Add a coverslip and examine the preparation for, and draw, fusion cells (zygotes) and asci.

Saccharomyces cerevisiae. Ascospore production in *S. cerevisiae* is achieved with some difficulty. *S. cerevisiae* possesses two somatic phases. An ascospore germinates to give a haploid cell which by budding produces a colony of smallish, round haploid cells. If two haploid populations of compatible mating strains ('a' and 'α') are mixed then some cell fusions occur to give diploid (not dikaryotic) cells. These are oval and larger than haploid cells and may continue to bud somatically producing diploid colonies until conditions suitable for sporulation occur. On sporulation a diploid cell becomes transformed directly into an ascus, and (usually) four haploid ascospores are produced after a meiotic division.

Streak a diploid culture of *S. cerevisiae* over a plate of sodium acetate agar (agar, 12 g; sodium acetate, 5 g; distilled water to 1 l) and incubate at 30 °C for 7 days. Scrape off some of the cells with an inoculating needle and smear them on a slide. Add a drop of water and a coverslip. Examine for ascospores. Contrast can be improved (with some loss of resolution in the image) by closing the iris diaphragm as far as is consistent with adequate light passing through the aperture. A better way to observe ascospores is to use a phase-contrast microscope or to stain differentially. This can be done as follows. Add a small drop of water to a clean (alcohol-washed) slide. Scrape some cells from the surface of the acetate agar and disperse them in the water. Spread the drop over the slide and allow the water to evaporate. Pass the slide fairly quickly three times through a bunsen flame, flood the smear with malachite green solution (1 per cent w/v in 1 per cent w/v aqueous phenol, prepared without heating). Heat to steaming (but do not allow the solution to boil) over a small flame for 2 min. Wash gently under the tap for 1 min, then flood with 0.5 per cent w/v aqueous safranin for 30 s. Wash again and examine with the high power objective (x 40). Ascospores retain the green stain; somatic cells and cell walls counterstain pink.

Anascosporogenous yeasts

Geotrichum candidum. Inoculate plates of 1 per cent malt agar with spores of *G. candidum* and incubate at 30 °C for 4 days. *G. candidum* is a mycelial yeast which reproduces by arthrospores. Cut out a block of agar from near the margin of a colony and examine the fungus without disturbing the mycelium.

Culture *G. candidum* in shake-flask conditions in 1 per cent malt broth. Mount a drop of the culture on a slide, apply a coverslip and examine this preparation.

Candida albicans. Candida is a genus showing well-marked dimorphism. It can exist as a budding yeast, particularly where the concentration of simple carbohydrates is high and the culture is well aerated, or it can exist as a mycelium producing blastospores. *C. albicans* is also characterized by its ability to produce chlamydospores.

Inoculate plates of 3 per cent malt agar and of cornmeal agar with *C. albicans.* Incubate for 7 days at 30 °C and compare the morphology of the fungus under the two conditions. *Cornmeal agar*: cornmeal (yellow), 40 g; agar, 15 g; distilled water to 1 l. Add cornmeal to boiling water (1 l) and simmer for 10 min. Filter through four layers of muslin into a 1 l graduated cylinder. Leave for 15 min and pour off the top 800 ml. Make up this 800 ml to 1 l, add the agar and sterilize.

Trichosporon cutaneum. Species of *Trichosporon* have a mycelium which produces both blastospores and arthrospores. Examine, using the technique as for *Geotrichum*, the morphology of a 1 per cent malt agar culture of this species.

Ballistosporogenous yeasts

There is one group of yeasts (called ballistosporogenous) which do not form ascospores but some of which can reproduce by ballistospores. A ballistospore is a small, asymmetrically-shaped spore which is borne on a fine peg (the sterigma). The ballistospore is forcibly liberated by a mechanism which depends on the secretion of a droplet of fluid from the spore at a point (the hilum) to one side of the junction of spore and sterigma. The spore and its liberation mechanism appear to be similar to those found in basidiomycetes (the spore, however, is not a basidiospore as there is no known nuclear fusion and meiotic process immediately preceding its formation. A technique for the isolation of ballistosporogenous yeasts is outlined in Flowsheet 22.

Tilletiopsis minor. Streak a 1 per cent malt agar plate with *T. minor.* Incubate the culture (inverted) for 7 days at 25 °C. Examine the lid for discharged spores; the liberation mechanism lifts the spores only some 0.5 mm above the colony, hence the inverted dishes. Examine the mycelium on the surface of the agar for sporulating areas and make observations on the secretion of the drop of fluid.

Sporobolomyces odorus. This is a non-mycelial, budding yeast. Streak cells of *S. odorus* on a plate of 1 per cent malt agar. As with *Tilletiopsis*, invert the plates during incubation. Examine the lid for discharged spores. Make a water mount of cells from the colony to observe budding. You may, in this preparation, see the occasional cell with a sterigma, but detection of sterigmata is made easier in a stained slide preparation. Proceed as follows. With the tip of an inoculating needle gently touch the *surface only* of a colony. The ballistospores are produced only on the surface of the colony, the more cells you include from the interior of the colony, the greater the difficulty in finding sterigmata. Disperse the cells picked up in a droplet of water on a clean slide. Spread the water and allow the slide to air-dry. Fix by passing the slide three times quickly through a bunsen flame. Allow to cool. Flood the slide with 1 per cent w/v methylene blue in 95 per cent ethanol for 3–4 s, then wash off thoroughly with running tap water. Examine the slide with the x 40 objective.

All species of *Sporobolomyces* are coloured due to carotenoid pigments. Ballistosporogenous, budding yeasts which are white and lack carotenoids are classed in the genus *Bullera.*

Rhodotorula rubra. This budding yeast is very similar to *Sporobolomyces.* Its species show a similar range of pigments and similar physiological properties. It does not, however, produce ballistospores. Examine *R. rubra.*

Flowsheet 22. Isolation of ballistosporogenous yeasts

Any old plant material is a potential source of the ballistosporogenous or 'shadow' yeasts. Collect samples of dead leaves, flowers, etc. The samples should be moist and not allowed to dry. Proceed as follows. Prepare several Petri dishes of cornmeal agar and, using Sellotape, attach some of the collected plant material to the lids of some of the dishes. Incubate the dishes at laboratory temperature (ideally 20 °C) and observe at daily intervals for the presence of small, pink colonies on the surface of the medium beneath the plant material.

↓

When colonies are observed the bottom of the plate should be placed upside down on top of another Petri dish bottom which also contains cornmeal agar. The two Petri dish halves are secured with Sellotape and left for approximately 1 h. Replace the lid on the second (lowermost) dish and incubate for 3 days to enable colonies to form from any spores discharged from the colonies on the first dish.

↓

This technique of purification can be repeated with a third plate or an isolated colony from the second plate can be suspended in sterile water (2 ml) in a test-tube and a loopful streaked across the surface of a solid medium. Transfer cells from purified colonies to slopes of malt agar.

Notes:

1. Cornmeal agar is preferred to malt agar for isolation as most of the ballistosporogenous yeasts eject more spores on a relatively poor growth medium such as cornmeal agar. An additional advantage of cornmeal agar is that most fungal contaminants which fall on the medium do not grow and spread as rapidly as on richer media such as malt agar.
2. From a pure culture, streak an inoculum on malt agar or cornmeal agar in a Petri dish. Invert the base of the dish over the base of a second dish of the same medium. Secure with Sellotape and incubate for 3 days. If the inoculum is streaked in a particular pattern on the first dish the same pattern will emerge on the second dish due to growth of spores discharged from the uppermost dish.
3. Examine young cultures and observe budding cells, sterigmata, and asymmetrical ballistospores. A convenient way to observe spores attached to sterigmata is to inoculate the particular species on a thin layer of cornmeal agar on a glass slide. Incubate the slide for 3 days on a V-shaped glass rod placed on a moist filter-paper in a Petri dish. Remove the slide and place the culture under the microscope; note the aerial sterigmata and attached spores.

4.3 BASIDIOMYCETES

There is a rich variety of fungi in this fungal class. The Basidiomycetes include mushrooms, toadstools, puffballs, stinkhorns, smuts, and rusts. Spores (basidiospores) are produced on the outside of a spore-producing body called the basidium. The basidiospores are usually uninucleate and haploid. They are formed as a result of plasmogamy, karyogamy, and meiosis, the last two processes occurring in the basidium. Four basidiospores are typically produced on each basidium.

The mycelium consists of septate hyphae which penetrate the substratum. In some species a number of hyphae lying parallel to one another are joined together to form thick strands of mycelium which are termed rhizomorphs. These can be observed in *Serpula lacrymans*, the fungus that causes dry-rot in timber, when grown on 1 per cent malt agar in glass Petri dishes or on slopes of the same medium.

The mycelium of most Basidiomycetes has three stages of development, the primary, the secondary, and the tertiary, during its life cycle. The primary mycelium develops from the germination of a basidiospore and may be multinucleate at first due to repeated division of the basidiospore nucleus or nuclei. This is usually a short stage in development as septa form and divide the mycelium into uninucleate cells. The secondary mycelium is commonly characterized by possessing clamp connections. The manner in which clamp connections form is illustrated in Fig. 26. They are usually interpreted as devices for ensuring the maintenance of the dikaryotic condition during apical growth and compartmentation, but a number of dikaryotic mycelia persist successfully without clamp connections. Some mycelia are produced with more than one clamp at each septum but this is superfluous to the dikaryotic maintenance needs. One possible function of clamp connections is the reduction of the impedance (resistance to flow) offered by the narrow opening of the dolipore septum which separates adjacent compartments. Clamp connections

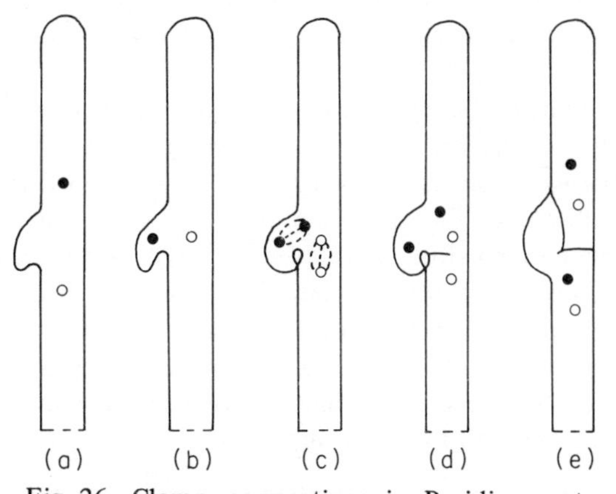

Fig. 26. Clamp connections in Basidiomycetes. Clamp connections are short hyphal branches which form during the period (a, b, c) that mitosis occurs in the dikaryotic nuclei. The clamp connection bypasses the septum which forms during cell division and allows one of the daughter nuclei formed during mitosis to migrate to the newly-formed subapical cell to which the complementary nucleus has already migrated. A second septum completes the separation of the two cells (d, e)

can be seen in 1-week-old cultures of *Serpula lacrymans* if some of the mycelium is stained with an ammoniacal stain such as congo red; the culture should be grown on 1 per cent malt agar and incubated at 25 °C.

The tertiary mycelium is represented by the specialized tissues which compose the sporophores (fruit bodies) of the higher Basidiomycetes. The cells of the tertiary mycelium are binucleate; the sporophores originate when the secondary mycelium differentiates to form complex tissue.

Lycoperdon perlatum. Species of *Lycoperdon* can be used to demonstrate spore liberation. Collect mature fruit bodies (puffballs) that have small, well-formed apical pores. If a fruit body is held beneath a source of dripping water, spores will be observed to puff out as each falling drop hits the peridium (the outer wall of the fruit body). Do not allow the water to drip on the apical pore or ostiole.

Agaricus bisporus. A. bisporus can be grown under laboratory conditions and used for several simple, but fundamental, experiments. One experiment consists of a study of the regions of expansion in the carpophore. This can be done as follows. With a fine pen draw a series of parallel lines 1 mm apart along the stalk (stipe) and cap when the mushroom is only about 2 cm high. The lines should be drawn at right angles to the long axis of the carpophore. A portion of the cap can be removed to expose the top of the stipe and this exposed region can also be marked. Measure the distances between the lines at daily intervals. The results, when graphed, should indicate that elongation of the stipe occurs primarily in the region immediately below the cap (Craig and Gull, 1977). Expansion of the cap is mainly at the margin.

In another experiment the relevance of the gills for stipe elongation can be demonstrated. Make two parallel cuts down through the centre of the cap and stipe, discard the material to the outside of each cut so that a T-shaped section of the cap and stipe remains. Remove the gills from one side of the cap and observe the subsequent behaviour of the fruit body. The stipe will curve away from the side bearing the gills. Use young fruit bodies (3 cm tall) for this experiment. The experiment can be modified in several ways. A similar curvature results if almost all of the cap tissue is removed but some of the gills are left attached to one side of the stipe. Also, curvature of a stipe can be induced when an aqueous extract of the gill tissue is applied unilaterally to a gill-free carpophore.

4.4 FUNGI IMPERFECTI

The Fungi Imperfecti are fungi in which the perfect stage, i.e. zygote, ascus or basidium, is lacking or not yet discovered. It should be stressed that the class is wholly artificial and is useful for the temporary reception of species pending discovery of structures showing that they belong either to the Phycomycetes, Ascomycetes, or Basidiomycetes. Although species are continually being removed from the Fungi Imperfecti the list is continually expanding because new species are being added more rapidly than old ones are being removed.

Almost all of the fungi assigned to the Fungi Imperfecti are probably Ascomycetes. The evidence for this includes the similarity between their conidial stages and those of Ascomycetes, the absence of unseptate mycelia characteristic of Phycomycetes, and the lack of clamp connections found in the Basidiomycetes. On the other hand, it cannot be assumed that there will be an eventual transfer of all species from the class because it is very likely that many species have lost the ability to form the perfect stage.

The system proposed by Saccardo has long been in use for classification of the imperfect fungi. The main basis of the system is the morphology of the mature sporulating structures; spore morphology and colour are also emphasized. The Saccardo system is very comprehensive in so much as it includes a place for all the genera so far described. More recently Hughes (1953) has proposed a more natural system for classification and this has been developed by other workers, including Barron (1968). The classification is based on the basic types of conidiophore and conidium development and has so far been applied to the Hyphomycetes (= Moniliales). Conidium and conidiophore morphology and colour are used as secondary characters for the purpose of classification.

It is proposed to limit study to the main ordinal group, the Moniliales. All the fungi described are readily grown on 1 per cent malt agar and the cultures should be incubated at 25 °C with the provision of some light. The method of inoculation is not critical and may be at one or more points on the medium. It should be remembered that cultures will usually remain usable longer when inoculation is near the edge of the dish. Use a sterile needle for inoculation and bring the needle upwards towards the inverted base of the plate. This technique tends to prevent surplus spores falling onto the medium.

Moniliales

Series Arthrosporae

The conidia are formed by septation and breaking up of simple or branched hyphae.

Geotrichum candidum. This organism is ubiquitous in soil and dairy products. It is saprophytic and sometimes pathogenic in human respiratory and gastrointestinal tracts. *G. candidum* has already been studied in several previous experiments. The conidia are one-celled and hyaline.

Series Meristem Arthrosporae

The conidia develop basipetally, originating from the meristematic apical region of the condidiophore. The youngest conidium in the chain hardly differs from the condiophore itself.

Trichothecium roseum. T. roseum causes a rot of apples. The colonies are white at first and then pink. The sporophores are erect and bear terminal clusters of spores directly attached to the tip. The spores are two-celled and roughly ovate with a nipple-like projection at the point of attachment.

Series Annellosporae

The first conidium forms as an apical aleuriospore; subsequent conidia form as swollen tips of conidiophores through the previous conidial scars. As a result the apical region of the conidiophore or sporogenous cell has a series of annellations.

Scopulariopsis brevicaulis. This species is found growing on all kinds of decomposing organic matter, particularly substrates containing a high percentage of protein, e.g. meat or cheese. The colonies are greyish-white at first, becoming yellowish-brown later. The conidia are lemon shaped and coarsely roughened.

Series Porosporae

Here the conidia develop through pores at the apex or side of the conidiophore. The conidiophores may be determinate in length or may continue growth through the apical pore or by sympodial extension from a lateral point near the apex.

Alternaria tenuis. This fungus is found on numerous kinds of organic materials. The spores are produced in chains, straight or branched, of up to a dozen or more. The spores are yellow-brown to dark-brown, very irregular in shape and size, with a definite beak. The spores are multicellular.

Series Blastosporae

Conidia develop as swollen ends of conidiophores or as buds from existing conidia. The conidiophores may be simple or branched and the conidia may be solitary or (more frequently) in simple or branched acropetal chains.

Cladosporium herbarum. This is a very common species which grows over a wide range of temperatures and frequently causes spoilage of meat in cold storage. It is also found on leather, paper, and textiles. The colonies are thick and velvety varying in colour from deep-green to dark grey-green. Under low magnification the spores are seen to occur in large, tree-like clusters. The sporing structures are very unstable and break up when mounted in fluid. The first-formed conidia increase by budding to form the tree-like masses. Younger spores are mostly one-celled, older ones are two- or even three-celled.

Series Botryoblastosporae

Conidia develop on swollen sporogenous cells that produce several conidia more or less simultaneously. The conidia may be solitary on conspicuous denticles (peg-like projections from the sporogenous cell) or may give rise to simple or branched acropetalous chains.

Botrytis cinerea. The name *Botrytis* derives from the Greek 'Botrys' = bunch of grapes. This describes the condidiophores which are irregularly branched at the top. The spores are borne on the ends of the final branches and each one is supported by a little peg or denticle (sometimes called a 'sterigma'). This fungus is found on a wide variety of materials and is parasitic on many kinds of plant. The colonies are pale brownish-grey. Look out for sclerotia which begin to develop after a few days as small, green, mycelial knots. These increase rapidly in size and eventually turn black.

Series Phialosporae

The conidia are cut off successively from the open growing point of or within the apical portion of the sporogenous cell (called a phialide). Conidia often remain in basipetal dry chains but sometimes are present in droplets of slime.

Aspergillus and *Penicillium.* The genera *Aspergillus* and *Penicillium* are probably the most commonly encountered of all fungi.

Aspergillus. This genus is widely distributed. The young mycelium produces an abundance of conidiophores which are not aggregated in any way but arise singly from the somatic hyphae. The hyphal cell which branches to give rise to the conidiophore is called a foot cell. The conidiophores are long, erect hyphae and each one terminates in a bulbous head, the vesicle. A number of phialides are produced from the surface of the vesicle, completely covering it. A layer of metulae may be produced, according to the species, between the phialides and the vesicle. The conidium-bearing phialides are bottle-shaped.

The conidia are formed inside the tip of the sterigma which is actually a tube. A portion of the protoplasm with a nucleus is delimited, the protoplast rounds off, secretes a wall of its own within the tubular sterigma, and develops into a conidium. The conidial wall may fuse partially or completely with the wall of the sterigma. A second protoplast below the first develops into a spore and pushes the first spore outwards without disjunction. In this manner a chain of spores is formed as the sterigma protoplasm continues to grow and cut off more conidia, one below the other.

Because conidiophores and conidia are produced in large numbers they give their predominant colour to the colony which they cover. *Aspergillus* colonies can be black, brown, yellow, green, etc. depending on the species and the medium on which the fungus is growing.

Examine different species of *Aspergillus* with the low power of your microscope. It is suggested that species with and without metulae are included. *Aspergillus janus*, the fungus used for an experiment in Flowsheet 23, has metulae whereas *Aspergillus giganteus* (also used in Flowsheet 23) has not. Press a strip of Sellotape over the colony and mount this over a drop of lactophenol + acid fuchsin on a slide. Make detailed drawings.

There are several species of *Aspergillus* which can be used to study the effects of

Flowsheet 23. Morphogenesis in *Aspergillus*

Experiment (1) *Effect of culture medium.* Inoculate plates of Czapek's solution agar, and the same medium to which 20 per cent sucrose has been incorporated, centrally with conidia of *Aspergillus repens.* Incubate at 25 °C. Note growth on the two media. Members of the *Aspergillus glaucus* group (to which *A. repens* belongs) are osmophilic and as a result growth will be poor on the first medium. On the medium rich in sucrose growth should be rapid and cleistothecia should be produced.

Experiment (2) *Effect of temperature.* Inoculate plates of Czapek's solution agar centrally with conidia of *Aspergillus janus.* Incubate separate plates at 20, 25, and 30 °C respectively for at least 12 days. Examine the morphology of the cultures. At 20 °C the conidial heads are white and the conidia are elliptical and smooth, the conidiophores are long with clavate vesicles. At 25 °C both green and white conidial heads may be present. At 30 °C the conidial heads are usually green and the conidia are globose and echinulate. The conidiophores are short with vesicles which are more rounded in shape than those produced at 20 °C. Hülle cells are also usually present at 30 °C.

Experiment (3) *Phototropism*. Prepare plates of solid media composed of one part of 1 per cent malt agar and one part of Czapek's solution agar. Inoculate plates centrally with conidia of *Aspergillus giganteus.* Incubate plates at laboratory temperature in small cardboard boxes with a slit cut in one side to provide unilateral illumination. If the plates are examined after 10 days the conidiophores will be seen to show orientation toward the slit.

Notes:
Czapek's solution agar: $NaNO_3$, 3 g; KH_2PO_4, 1 g; $MgSO_4 \cdot 7H_2O$, 0.5 g; KCl, 0.5 g; $FeSO_4 \cdot 7H_2O$, 0.01 g; sucrose, 30 g; agar, 15 g; distilled water to 1 l. When preparing the medium it is suggested that all the salts except the phosphate are dissolved in approximately half the water and the sugar added. Dissolve the phosphate separately and add to the rest, then make up to the required volume.

growth medium, temperature, and light on development. These experiments are simple to carry out and appropriate procedures are given in Flowsheet 23.

Penicillium. The Penicillia are as common and as cosmopolitan as the Aspergilli. They are the so-called green moulds and blue moulds that are frequently found on citrus and other fruits, on jellies and preserves, and other foodstuffs. They are frequent laboratory contaminants. Industrially, the Penicillia are important in making cheese and antibiotics. *Penicillium roqueforti* is responsible for the flavour of Roquefort cheese and *Penicillium camemberti* for that of Camembert cheese *Penicillium notatum* and *Penicillium chrysogenum* are sources of the antibiotic penicillin (see Flowsheet 24). Other species produce this antibiotic but certain selected strains of *P. chrysogenum* have proved to be most efficient for its manufacture.

Flowsheet 24. Demonstration of antibacterial substances produced by species of *Penicillium*

Prepare plates of the following medium: sucrose, 30 g; glucose, 2 g; peptone, 3 g; yeast extract, 2 g; $NaNO_3$, 2 g; K_2HPO_4, 1 g; $MgSO_4 \cdot 7H_2O$, 0.5 g; KCl, 0.2 g; $FeSO_4 \cdot 7H_2O$, 0.01 g; agar, 15 g; distilled water to 1 l. Adjust the medium to pH 7.0 before sterilization. After sterilization pour 15 to 20 ml in each of several Petri dishes. Allow the plates to dry by leaving them in an incubator at 35 °C for 1 day before use. Streak spores of *Penicillium chrysogenum* in a straight line (approximately 5 cm long) close to the side of the dish. Repeat for other species of *Penicillium*. Incubate the cultures for 3 days at 25 °C.

Prepare cultures of *Staphylococcus aureus, Bacillus subtilis* and *Bacillus cereus* by streaking cells of each bacterium on separate slopes of nutrient agar. Incubate at 30 °C for 2 days.

↓

Add 5 ml sterile distilled water to each slope culture, shake, and streak a loopful of cells from each slope at right angles to each colony of *Penicillium*. Incubate the plates for a further 24 h at 30 °C and examine for evidence of antibiotic production by the fungi used.

Notes:
1. *Staphylococcus aureus* and *Bacillus subtilis* are sensitive to penicillin and growth of these bacteria should be inhibited at the ends of the appropriate streaks nearer the fungus colony – assuming the fungus produces penicillin.
2. *Bacillus cereus* produces a penicillinase and should grow almost right up to the margin of each species of *Penicillium* tested.

The life history of a typical *Penicillium* is very much like that of *Aspergillus* but the morphology of the reproductive structures differs considerably. The mycelium produces septate, long, erect conidiophores which branch about two-thirds of the way to the tip in a characteristic symmetrical or asymmetrical broom-like fashion. The conidiophore, often referred to as the brush, is technically known as the penicillus. The branching of the conidiophore ends in a group of phialides which bear the long conidial chains. In penicilli with more than one stage of branching the branches bearing the phialides are, as in *Aspergillus*, called metulae. The branches supporting the metulae, if comparatively short and obviously part of the penicillus, are known as rami. The conidia are globose to ovoid and resemble glass beads when viewed under the microscope. As in the *Aspergilli*, the enormous quantities of conidia produced are chiefly responsible for the characteristic colony colour of various species of *Penicillium*.

The large number of species of *Penicillium* fall naturally into three main sections:

Monoverticillata: the penicilli consist of a single whorl of phialides.

Biverticillata-symmetrica: each species has a compact whorl of metulae (= branches), each bearing phialides, the whole penicillus being approximately symmetrical about the axis.

Asymmetrica: includes all species in which the penicillus is branched more than once and is asymmetrical about the axis, or, if approximately symmetrical, has not the compact structure or tapering phialides of the members of Biverticillata-symmetrica.

Culture species of each of these main series of *Penicillium* on 1 per cent malt agar and examine the structurely different types of penicilli. Suggested species for study are:

Penicillium frequentans (Monoverticillata)

Penicillium purpurogenum (Biverticillata-symmetrica)

Penicillium chrysogenum }
Penicillium digitatum } (Asymmetrica)

Mycelia Sterilia

Conidia are not produced. Bulbils (irregular clusters of cells) and sclerotia may be produced.

Melancoliales

Acervuli usually form on host in nature. In culture the conidiophores may be separate, clustered into flat or cushion-shaped masses, or enclosed in sterile tissue.

Sphaeropsidales

Conidia produced in pycnidia that may develop singly or in groups. Conidia are usually well formed in culture.

5
Nutrition of Fungi

Fungi are heterotrophic for carbon compounds and these serve two essential functions in fungal metabolism. The first function is to supply the carbon needed for the synthesis of compounds which comprise the living cell. Proteins, nucleic acids, reserve foods, etc. would be included here. Second, the oxidation of carbon compounds produces appreciable amounts of energy. Fungi can utilize a wide range of carbon sources such as monosaccharides, oligosaccharides, polysaccharides, organic acids, and lipids. Carbon dioxide can be fixed by some fungi but cannot be used as an exclusive source of carbon for metabolism. The effect of different carbon compounds on growth rate is readily determined and Flowsheet 25 gives details of an experiment to test the effect of glucose, starch, and sodium acetate on the radial growth of fungal colonies.

Different strains of a fungus can vary in the ability to utilize certain carbon compounds. Also, when presented with a mixture of carbon compounds, there may be a preferential utilization of one form over others by particular fungi. This situation occurs in *Penicillium chrysogenum* when supplied with a mixture of acetate, lactate, glucose, and lactose. Acetate is the first carbon source to be utilized, lactose the last. In some cases utilization of sugars is more effective when mixtures, rather than single sugars, are supplied. An example is *Aspergillus niger* which shows a marked increase in dry weight when cultured in liquid media containing both galactose and glucose, dry weight yields on either galactose or glucose are much lower (Horr, 1936).

Most fungi utilize inorganic or organic sources of nitrogen. Some fungi are able to utilize both nitrate and ammonia, others can use ammonia but not nitrate. Many fungi can utilize amino acids as a source of nitrogen. When ammonium salts such as the sulphate, nitrate, or chloride are added to media there is often a marked drop in pH as a result of the preferential utilization of the ammonium cation. Growth can be reduced as a result of this effect.

It is known that fungi have relatively large requirements for magnesium, phosphorus, potassium and sulphur. Magnesium is usually supplied as the sulphate and its main function appears to be the activation of enzymes necessary for metabolism and growth. Phosphorus, usually provided as potassium phosphate, is absorbed rapidly during growth; deficiency results in a reduced rate of glucose utilization. The rate of absorption of phosphate is reduced when growth becomes limited by the availability of either carbon or nitrogen. A deficiency in potassium

Flowsheet 25. The effect of different carbon sources on fungal growth

---

Prepare separate batches of Czapek's solution agar (see Flowsheet 23) containing 1 per cent w/v glucose, 1 per cent w/v starch, 1 per cent w/v sodium acetate, and no added carbon source. Pour several plates of each medium and allow to dry overnight at 35 °C. Inoculate the centre of each plate with spores from a 7-day-old malt agar slope culture of *Penicillium chrysogenum*. Incubate all the cultures at 25 °C.

↓

After 2 days, and then at 1-day intervals, measure the diameter of the colonies which form on the various media. Use the measurement technique described in Section 2.2. Plot a graph of colony diameter against incubation time for each of the different carbon sources. Make notes on the morphology of the cultures.

---

results in poor sugar utilization, and in *A. niger* a reduction in potassium results in the accumulation of oxalic acid. Sodium can partially replace potassium in *A. niger.* Most media contain sulphur as magnesium sulphate and most fungi can supply their needs from inorganic sulphate – the sulphate is reduced and incorporated into organic molecules. Flowsheet 26 contains details of an experiment to investigate mineral deficiencies in *A. niger.*

Fungi have a requirement for the following micronutrients: iron, zinc, copper, manganese and molybdenum. Calcium is also required by fungi but the requirement is of the same order as a micronutrient. Fungi require iron at a concentration of approximately 0.1 mg l^{-1} in culture media. At neutral or alkaline pH iron is lost from solution and it is useful to include a chelating agent. There are many important metabolites which contain iron: examples are catalase, cytochromes, and the ferrichromes. Certain metabolites, particularly antibiotics such as penicillin and streptomycin, are produced in maximum concentration when the iron concentration is greater than that required for maximum growth of the culture concerned. Zinc is essential for fungal growth; a concentration of 1 mg l^{-1} is considered adequate for routine work. The metabolic effects of zinc deficiency are numerous and probably result from the role of the metal in the synthesis of enzyme proteins. Copper is required for normal growth and sporulation at a concentration of approximately 0.1 mg l^{-1}. One of the most obvious effects of copper deficiency is a decrease in pigmentation of spores of *Trichoderma viride* and *Aspergillus* spp. Manganese is essential at levels of about 0.01 mg l^{-1} and deficiency results in decreased sporulation. Molybdenum is required in very small quantities (from 0.1 to 10 mg l^{-1}). Calcium has optima reported in the range 0.5 to 20 mg l^{-1} and species from the major groups of fungi respond to added calcium by an increase in dry weight. Some fungi, however, do not appear to have a requirement for calcium.

Fungi require vitamins in addition to the nutrients and micronutrients which have already been discussed. It has been assumed that a fungus which grows in the

Flowsheet 26. Mineral deficiencies in cultures of *Aspergillus niger*

Prepare 1 l of each of the following solutions:

Solution No. 1 Glucose, 40 g l^{-1}
2 Magnesium chloride ($MgCl_2 \cdot 6H_2O$), 1.25 g l^{-1}
3 Magnesium sulphate ($MgSO_4 \cdot 7H_2O$), 1.25 g l^{-1}
4 Potassium chloride (KCl), 5 g l^{-1}
5 Potassium dihydrogen orthophosphate (KH_2PO_4), 2.5 g l^{-1}
6 Potassium nitrate (KNO_3), 5 g l^{-1}
7 Potassium sulphate (K_2SO_4), 2.5 g l^{-1}
8 Sodium dihydrogen orthophosphate (NaH_2,PO_4), 5 g l^{-1}
9 Sodium nitrate ($NaNO_3$), 5 g l^{-1}
10 Sodium sulphate (Na_2SO_4), 1.25 g l^{-1}

Prepare 160 ml of each of the following media by mixing 40 ml aliquots of the appropriate solutions in a medical flat bottle:

	Medium	*Nutrient status*
Solutions No.	1, 3, 5, and 6	Complete
	1, 5, 6, and 10	Minus magnesium
	1, 3, 4, and 5	Minus nitrogen
	1, 3, 6, and 7	Minus phosphate
	1, 3, 8, and 9	Minus potassium
	1, 2, 5, and 6	Minus sulphur

Add 2 g of agar to each bottle and sterilize.

↓

Pour plates of each medium and dry plates overnight at 35 °C. Inoculate the centre of the medium on each plate with spores from a 7-day-old malt agar slope culture of *A. niger*. Incubate at 30 °C.

↓

Record the diameter of each colony after 2 days and then at daily intervals. Use the measurement technique described in Section 2.2 and draw a graph of colony diameter against time for each treatment.

Notes:
1. Use 'Analar' grade reagents for this experiment as 'general purpose' reagents may contain significant levels of other elements which are not required.
2. Examine the morphology of *A. niger* under each of the experimental conditions. Note any effects on sporulation and spore colour.

Flowsheet 27. The production of biotin by *Aspergillus rugulosus* and the demonstration of biotin requirement by *Sordaria fimicola*

Prepare two plates of the following medium: glucose, 5 g; magnesium sulphate, 0.75 g; potassium dihydrogen orthophosphate, 1.75 g; potassium nitrate, 3.5 g; agar, 12 g; distilled water to 1 l. Inoculate one plate with spores from a 10-day-old malt agar slope culture of *S. fimicola.* Inoculate *S. fimicola* at one side of the other plate and inoculate spores from a 7-day-old malt agar slope culture of *A. rugulosus* on the opposite side. Incubate the plates at laboratory temperature in the light for 10 days.

↓

Examine the plates carefully. Are there any perithecia (small, black spheres) produced by *S. fimicola* on the plates? If present, where are they found in relation to the growth of *A. rugulosus*?

Notes:
Remove one or more perithecia, place on a slide and add a drop of lactophenol. Press gently with a needle to release the asci. Count the spores in some of the asci which are extruded.

absence of an exogenous supply of a vitamin does so as a result of its ability to synthesize the compound. Vitamins play a catalytic role in cell metabolism; they are also metabolites and are synthesized and destroyed in the cell. The culture medium of a fungus usually contains any vitamins which are synthesized by the fungus and these can be detected by bioassay. There are exceptions and in some fungi the greater part of the vitamins remain in the cells. Thiamine (aneurin or vitamin B_1) is the vitamin most frequently required by fungi. Biotin (vitamin H, coenzyme R) is required by many yeasts and fungi. Pyridoxine (vitamin B_6, adermin) is required by fewer fungi than is biotin or thiamine. Other vitamins include riboflavin (vitamin B_2, lactoflavin), which is only rarely required by microorganisms isolated from nature, inositol, nicotinic acid, *p*-aminobenzoic acid (PABA), pantothenic acid, and vitamin B_{12}. Flowsheet 27 gives details of an experiment to study the effect of biotin deficiency on *Sordaria fimicola. S. fimicola* does not normally produce perithecia unless the vitamin is supplied and biotin deficiency results in failure of the ascospores to mature (Barnett and Lilly, 1947). The former observation is the basis of the experiment in Flowsheet 27 which is based on an experiment by Dade and Gunnell (1969).

6
Morphogenetic Substances Produced by Fungi

Many, if not all, fungi produce metabolites which can retard growth or induce morphogenetic effects when applied to actively-growing hyphae of the same or other species. Metabolites which induce branching, bulging, narrowing, or vacuolation of hyphae have been described. The possible relevance of such metabolites to fungal morphogenesis is discussed in Chapter 7. In this chapter it is proposed to concentrate on metabolites which induce vacuolation as these can be readily demonstrated.

Culture filtrates of *Fusarium oxysporum* and many other fungi contain metabolites which induce the actively-growing hyphal tips of several species of fungi to vacuolate. It is beyond the scope of this book to describe the fine structure of fungi in detail but it should be mentioned that the hyphal-tip zone of an actively-growing hypha is normally non-vacuolate. There are in fact minute vacuoles present but these cannot be observed with light microscopy. When leader hyphae of a fungal colony are treated with a culture filtrate which contains these so-called 'vacuolation factors' a sequence of events occurs which results first of all in a cessation of extension-growth. This is followed by the development of vacuoles in the hyphal-tip region. The higher the concentration of the applied vacuolation factors the more rapid the cessation of extension-growth and the greater the degree of vacuolation. At low concentrations of the factors just the extreme tip of the apical cell may vacuolate but vacuolation may spread to affect the whole apical cell when high concentrations of the factors are applied. Metabolites with vacuolation-inducing activity have also been isolated from various microbial habitats such as soil and pond water (Park and Robinson, 1964).

Fusarium oxysporum produces at least two vacuolation factors. One of these factors has been isolated and characterized as bikaverin (Cornforth *et al.*, 1971). The extraction procedure and formula for bikaverin are shown in Fig. 27. This metabolite is extremely active as a vacuolation factor; a solution of 0.01 $\mu g\ ml^{-1}$ induces vacuolation in the test organism (*Aspergillus niger*). If the procedures outlined in Flowsheet 28 are followed it is a straightforward matter to demonstrate these vacuolation factors.

As fungal cells age they become more vacuolate such that there is a polarity of vacuolation in each hypha (Fig. 28). There is also a polarity of transport,

Flowsheet 28. Fungal metabolites which induce vacuolation in apical cells of hyphae

---

Inoculate 50 ml sterile S liquid in a 500 ml conical flask with *Fusarium oxysporum*. The flask is plugged with cotton wool. Incubate for 7 days at 25 °C.

↓

Filter the culture by pouring through a nylon strainer or filter-paper. It is not essential to remove all fine particles, propagules, etc. to demonstrate the action of the culture filtrate. Determine the pH of the culture filtrate and adjust to pH 3.5 with 0.1 M-HCl or 0.1 M-NaOH. Use the culture filtrate for the following experiments.

Experiment (1) Prepare serial dilutions of the culture filtrate in distilled water (adjusted to pH 3.5) to give the following dilutions: 1/2, 1/4, 1/8, 1/16, 1/32, 1/64. Apply *one* drop (approximately 0.02 ml) of each dilution to overlap a portion of the margin of colonies of *Aspergillus niger* and *F. oxysporum* inoculated as diametric streaks on S agar and incubated for 3 days at 25 °C prior to treatment. Record the presence or absence of vacuolation of the hyphal tips 10 min after application of the dilutions of the filtrate. Determine the vacuolation value (VV) of the culture filtrate; this is defined as the highest dilution which can induce vacuolation. If the 1/8 dilution causes vacuolation but the 1/16 dilution is not effective the VV is between 8 and 16 and is recorded as 8.

Experiment (2) Adjust separate portions of the culture filtrate to pH 3, 4, 5, and 6. Prepare serial dilutions (up to 1/64) of each portion in distilled water of the same pH. Assay each dilution as described in Experiment (1) but only use *A. niger* as the test organism. How does the activity vary with the pH of the applied solution?

Experiment (3) Boil a known volume of culture filtrate for 5 min. Adjust to the original volume with distilled water and adjust the pH to 3.5. Dilute and assay as in Experiment (1) but only use *A. niger* as the test organism. What effect does boiling have? Use an unboiled sample of culture filtrate for comparison.

Notes:

1. When preparing serial dilutions in distilled water it is convenient to add 2 ml aliquots of distilled water to a series of test-tubes. A 2 ml sample of culture filtrate is added to the first tube, mixed, and 2 ml transferred to the second tube and so on until the requisite dilutions are made. Take care either to change or carefully rinse the pipette in clean water between transfers to the test-tubes. If a Pasteur pipette is used to apply drops to the colony margin, the same one can be used if the most dilute sample of culture filtrate is applied first.
2. It is convenient to indicate the intended areas of application on the margin of the culture by means of a circle on the base of the culture dish. It is recommended that six drops are applied to one colony, three to each of the parallel margins. Avoid applying drops to the colony margin adjacent to the edge of the Petri dish.
3. As an extension of these experiments, study what happens to hyphae which have been induced to vacuolate. Do the vacuoles eventually regress? Do the hyphae continue growth? If so, how?
4. Different strains of *F. oxysporum* produce different quantities of vacuolation factors and it may be useful to try more than one strain in these experiments. Although all the strains tested at Belfast produced vacuolation factors it is of course possible that there are strains which do not.

Fusarium oxysporum was grown at 22 °C in the light in distilled water (20 l) containing glucose, 200 g; $MgSo_4 \cdot 7H_2O$, 10 g; KH_2PO_4, 4 g; and NH_4NO_3, 2 g. The culture was aerated. After 7–10 days the mycelium was collected on muslin, washed with water, dried in a current of cold air, ground finely, and extracted for 24 h in a Soxhlet apparatus with light petroleum (b.p. 40–60 °C). The evaporated extract showed no vacuolation activity.

↓

The mycelial powder was then extracted with chloroform for 24 h. The deep-red extract was filtered and reduced in volume on a steam-bath. The residue was triturated several times with light petroleum ether and then collected on a filter and dried. Yields of the crude pigment ranged from 4 to 8 mg g^{-1} of dried mycelium.

↓

The crude pigment (bikaverin) was recrystallized from chloroform and dried at 140° C *in vacuo*. It formed deep-maroon laths with the following structure:

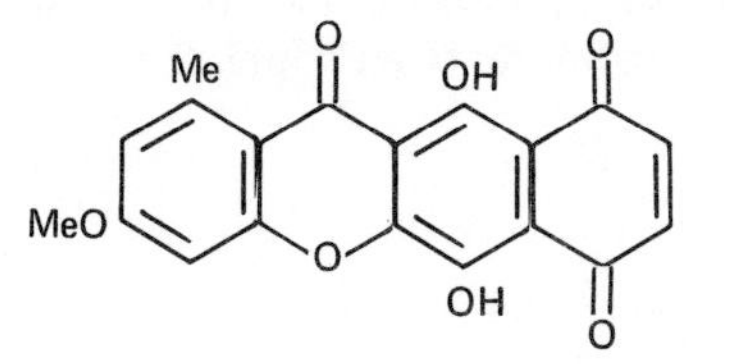

For bioassay, the pigment was dissolved either in dimethyl formamide or in refluxing acetone and the solutions were diluted with relatively large volumes of water. Vacuolation was observed when the solution was assayed at a concentration of 0.01 mg l^{-1} (~3×10^{-8} M) on marginal hyphae of colonies of *Aspergillus niger*. (Cornforth *et al*., 1971)

Fig. 27. Extraction procedure for bikaverin

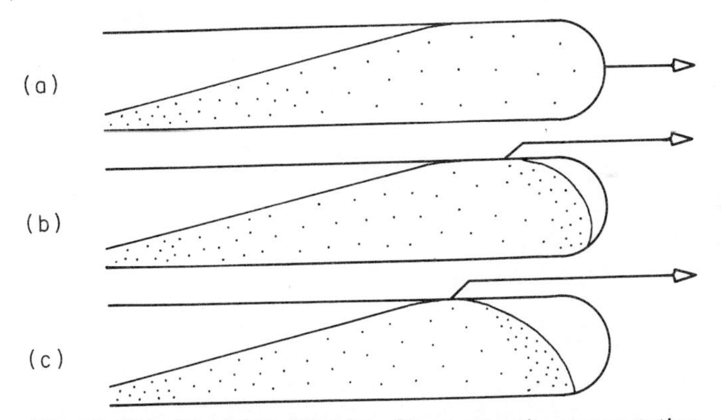

Fig. 28. Polarity of fungal hyphae. Diagrammatic representation of a hypha of *Aspergillus niger* before and after treatment with vacuolation factor. Stippled area = volume and density of cytoplasmic matrix; white area = vacuolar volume. Arrows indicate origin of subsequent growth. (a) Normal tip with apical extension-growth, the cytoplasm is least dense at the tip; (b) low concentration of vacuolation factor causes vacuolation immediately behind the apex, cessation of apical extension-growth, and subsequent subapical branching; (c) higher concentration of vacuolation factor causes greater degree of vacuolation near tip. Subsequent growth is by lateral branching proximal to the induced vacuoles (Park and Robinson, 1967). Reproduced by permission of Oxford University Press

metabolites moving distally in the hyphae (Park and Robinson, 1967). It is very tempting to consider the possibility that these three polarities (age, transport, and vacuolation) are more than coincidentally linked. With this in mind hypotheses have been formulated to link vacuolation with ageing and transport. Vacuolation factors produce vacuoles which are identical to those forming naturally with age and the idea that these or similar metabolites may be ageing hormones has been explored. Vacuolation of hyphal-tip cells is a common feature of an old, staled culture in a liquid or on a solid medium but there is no good evidence to suppose that the factors so far studied play any role in natural ageing.

The vacuolation factors produced in liquid culture by *F. oxysporum* are present at maximum concentration during the later part of the exponential growth phase. This is true of many other metabolites (see Chapter 7).

7
Staling and Colony Morphology

Staling in fungi can be considered as a retardation or cessation of growth in a culture containing a fixed quantity of nutrients. Numerous hypotheses have been put forward to account for staling but the most credible are that staling may result from a decline in concentration of one or more nutrients essential for growth or from an increase in concentration of one or more growth-inhibitory metabolites. Staling could, of course, result from a combination of these two factors.

Staling has been described for several culture conditions: in colonies growing on solid media, in shake-flask culture, and in continuous culture. These growth conditions differ markedly and it is useful to look at each situation briefly if only to illustrate the complexity of staling as a whole.

Colonies on solid media

When a fungus is inoculated on a solid medium in a Petri dish the colony diameter usually increases at a uniform rate until the marginal or leader hyphae approach the margin of the dish. It was noted by Brown (1925) that some strains of *Fusarium* species began to decrease in radial growth rate well before they reached the perimeter of certain media. The final colony diameter varied with the concentration of the growth medium (see Fig. 29). Although the radial growth rate declined and ultimately ceased there was an increase in the intensity of aerial mycelium until it extended to the staling colony margin. As a result, a decline in radial growth rate need not be reflected in a decrease in the mass of the colony as a whole. This seems to be a localized type of staling in so much as only the extension-growth of the marginal hyphae is affected.

There is a more general type of staling which occurs in colonies on solid media. This is particularly clear in *Fusarium oxysporum* (Park, 1961) and results in the development of a predictable pattern of differentiation from the actively-growing colony margin to the centre of the colony. This general type of staling was the subject of the practical work in Section 3.1.

Shake-flask culture

Here the situation is somewhat simplified in the sense that any gradients in either nutrient or toxic metabolite concentrations are unlikely to be maintained in

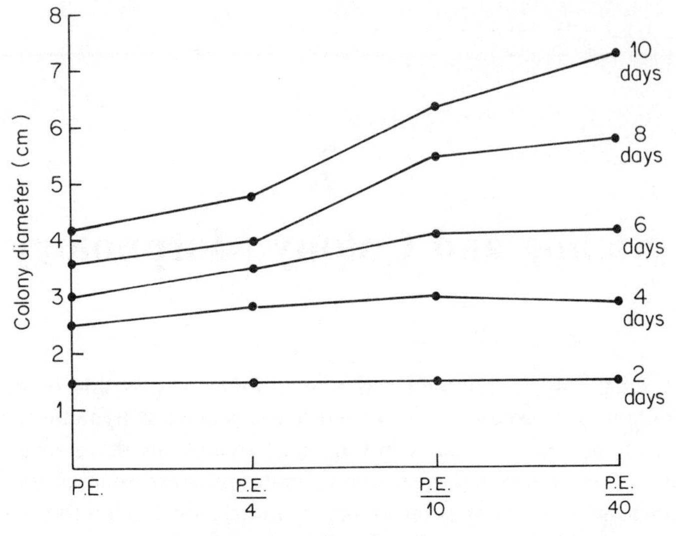

Fig. 29. Growth of a strain of *Fusarium* on various concentrations of potato extract agar (P.E.). The colony diameter was recorded at 2-day intervals on each medium. A marked falling off in the growth rate from about the fourth day occurred on the two higher concentrations of P.E. medium used. Staling did not occur at the two lower concentrations of P.E. medium studied (Brown, 1925). Reproduced by permission of Oxford University Press

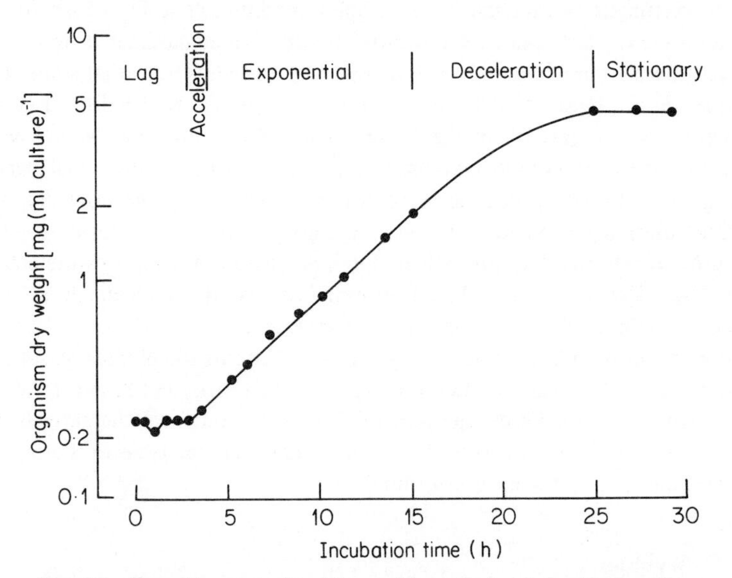

Fig. 30. Growth curve for *Mucor hiemalis* in shake-flask culture. Reproduced, with permission, from Trinci, *Transactions of the British Mycological Society*, **58**, 467–473 (1972). Published by Cambridge University Press.

relation to the culture as is the case on solid media. This results in a more uniform environment. In liquid cultures growth is normally estimated on a dry weight basis and staling is considered to commence at the beginning of the stationary phase of growth or slightly earlier at the deceleration phase in specific growth rate (Fig. 30). The deceleration phase occurs prior to the stationary phase and is more marked in some species than others. Spore formation can commence during the exponential phase of growth in some species grown in media in which growth is limited by the concentration of glucose. It normally commences, however, during the deceleration or stationary growth phases. For some fungi, increasing the initial glucose concentration of the medium can result in the onset of staling at a lower dry weight than at lower glucose concentrations. This could be due to oxygen limitation or an accumulation of toxic metabolites.

Continuous culture

Here we have the best clues as to the mechanism of staling. The work of Righelato *et al.* (1968) has established the role of glucose in differentiation in glucose-limited chemostat cultures of *Penicillium chrysogenum*. When growth is stopped by cutting off the supply of medium to the culture and glucose is supplied to the culture at the maintenance ration (the ration necessary to keep the existing dry weight of the culture constant) there is a marked increase in hyphal vacuolation. Even the hyphal-tip regions vacuolate. Phialides (cells which produce conidia) and conidia appear; the phialides are usually formed singly at the tips of hyphae. Autolysis results when the glucose feed is shut off completely. Continuous culture indicates that the cessation of growth and the differentiation seen behind the margin of a colony on a solid medium can be correlated with the decline in concentration of one or more energy-providing substrates in this region.

What evidence is there that staling substances are produced which contribute to or cause some of the symptoms which define staling? Many growth-inhibitory metabolites produced by fungi in liquid culture have been characterized and occur at maximum concentration during the deceleration phase of growth. The growth-inhibitory (and morphogenetic) activity of such metabolites has often been demonstrated by applying some culture filtrate or extract of the filtrate to a test organism growing on solid medium (see Chapter 6). It is one thing to demonstrate such effects with extracts or culture filtrates applied to a test organism, quite another to determine the role of such metabolites in the cultures from which they were isolated. Also, many experimenters purporting to demonstrate morphogenetic effects due to staling metabolites have not completely eliminated the possibility that the effects could be explained by nutrient depletion in the assay system.

The question of tolerance is relevant. Growth-inhibitory metabolites and morphogenetic factors produced by liquid cultures of a fungus are frequently assayed on the leader hyphae of colonies (often of a different species) growing actively on solid media. The effects are often more dramatic. However, the assay organism has not necessarily been growing in the presence of the metabolites being tested and is often growing rapidly at the time of assay. This point is particularly relevant since the reaction of hyphae to growth-inhibitory metabolites can be

Table 2. Relationship of colony age and sensitivity of hypal tips of *Aspergillus niger* to vacuolation factor. Reproduced, with permission, from Robinson and Park, *Transactions of the British Mycological Society*, **48**, 561–571 (1965). Published by Cambridge University Press

Days from inoculation of *A. niger* on S agar	1	2	3	5	9
Colony width (mm)	1	5	8	24	50
Sensitivity[a]	0.19	0.25	1	1	1

[a]A high value corresponds to a high sensitivity to vacuolation factor.

governed by the age, diameter, and rate of extension-growth of the hyphae (Table 2). Young, narrow, and slow-growing hyphae may be particularly resistant. Staling is a function of declining growth and although it could be argued that many metabolites could contribute to this condition, the hyphae when growing more slowly are likely to be more resistant to their own growth-inhibitory metabolites and to certain substances which can induce morphogenetic effects in test fungi. The numerous growth-inhibitory (when assayed) metabolites which peak in concentration during the late exponential or deceleration phase in liquid cultures may merely reflect the greater metabolic activity of the fungus during the exponential growth phase (Fig. 31). There is no direct evidence that one or more of these metabolites initiate staling.

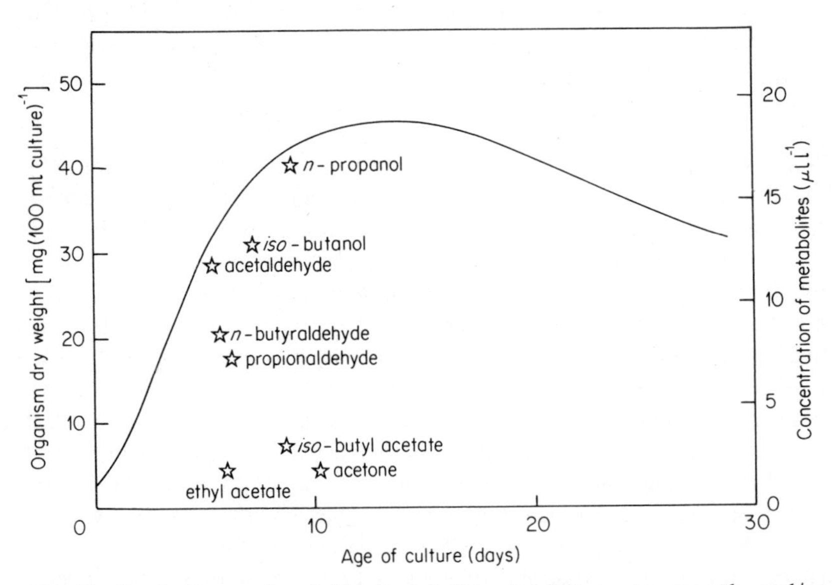

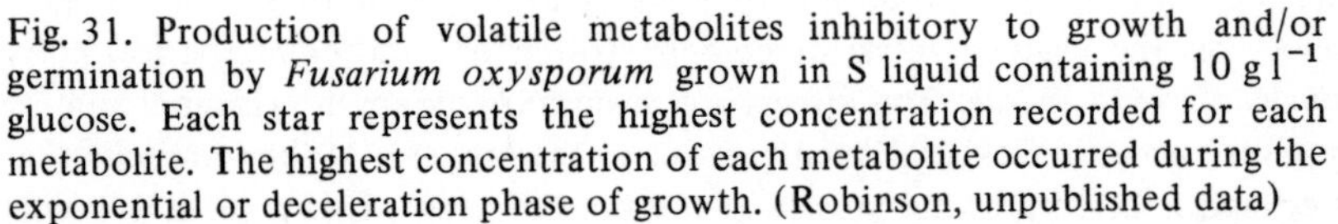
Fig. 31. Production of volatile metabolites inhibitory to growth and/or germination by *Fusarium oxysporum* grown in S liquid containing 10 g l^{-1} glucose. Each star represents the highest concentration recorded for each metabolite. The highest concentration of each metabolite occurred during the exponential or deceleration phase of growth. (Robinson, unpublished data)

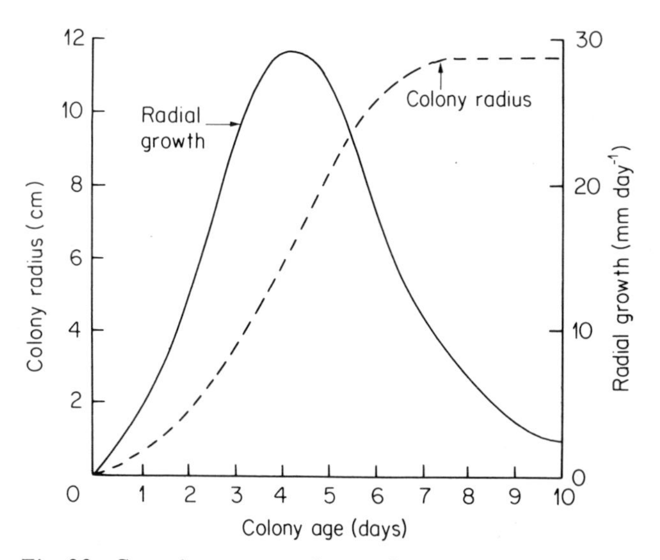

Fig. 32. Growth pattern of a surface culture of *Rhizoctonia solani*. The culture was suspended on the surface of a liquid medium (20 l) at the centre of a large glass cylindrical growth container. The fungus ceased growing after 8–9 days by which time the colony was within 2–3 cm of the wall of the container. The growth restriction was not caused by the wall of the container because when containers of smaller diameter were used the hyphae not only reached the sides but even grew up the wall. (Gottlieb, 1971). Reproduced by permission of The New York Botanical Garden

While on the subject of the existence or non-existence of staling substances, Gottlieb (1971) describes experiments in which radial growth of *Rhizoctonia solani* is limited when colonies are grown on the surface of liquid media (Fig. 32). There appears to be a limit to the diameter the colony can attain and Gottlieb postulates the production of an endotoxin which limits growth. The evidence presented is by no means conclusive.

Staling is not the overall cessation of growth so often imagined. It is a dynamic process and the resultant morphological changes are not synchronized. Whereas many hyphae in staled liquid cultures may vacuolate, some continue to grow normally, some autolyse, and others differentiate to form propagules. On solid media, hyphae may lyse after they have produced propagules; the propagules may subsequently germinate *in situ* and the germ-tubes lyse later. This is the 'cryptic growth' which was the subject of an earlier exercise (Section 3.4).

Experiments to demonstrate two aspects of staling are outlined in Flowsheet 29. Experiment (1) demonstrates that liquid from a staled culture can support spore germination and hyphal growth. Initially the drops of staled culture liquid will contain non-germinated conidia and possibly chlamydospores. Some hyphae will also be present but many of these will be vacuolate at the tips and others will be

Flowsheet 29. Staling in fungi

Experiment (1) Sterilize 20 ml batches of S liquid medium in 100 ml wide-necked flasks plugged with cotton wool. Inoculate each flask with 0.5 ml from a spore suspension (5 ml) prepared from a 7-day-old slope of *Fusarium oxysporum* grown on malt agar at 25 °C. Incubate flask cultures at 25 °C for 20 days or longer.

↓

With a sterile Pasteur pipette transfer drops of staled liquid from these cultures to sterile glass slides in humid chambers. Examine some of the drops immediately and others after incubation for 24 h at 25 °C.

Notes:
1. When transferring staled liquid take care to sample the culture in a region fairly free of visible mycelium. Some propagules and hyphae will still be included in the liquid and serve as an inoculum.
2. A suitable humid chamber can be made from a Petri dish lined with moist filter-paper. Two glass rods or a V-shaped glass rod placed on the filter-paper serve as a support for one or two sterile slides.
3. As an alternative or additional experiment, remove the mass of the mycelium from a 20-day culture of *F. oxysporum* and incubate the solution at 25 °C. Take samples from the solution at daily intervals over a period of 3 days. Make notes on the development of the indigenous conidia and chlamydospores.

Experiment (2) Prepare four Petri dishes of S agar (15 ml) and inoculate all the plates with a small central inoculum using one of the following species: *Rhizopus stolonifer, Rhizopus arrhizus, Cunninghamella elegans*, or *Geotrichum candidum*. Incubate cultures at 25 °C.

↓

After incubation for 1, 2, 4, and 6 days (in the case of *G. candidum*) or 1, 2, 3, and 4 days for each of the other species, remove one plate and place a Cellophane strip over the colony. Streak spores of the same species over the Cellophane and record germination after 3 to 4 h and 24 h in a zone over the colony centre. Seed spores on Cellophane over sterile S agar as a control.

Notes:
1. Prepare the Cellophane as in Flowsheet 1. Strips 8 x 2 cm are a useful size and are placed diametrically on the colonies.
2. It is recommended that all the cultures used to provide spores for inoculation on the Cellophane are grown on S agar for 5 days at 25 °C. Sterile water (5 ml) is added to each slope which is shaken gently and the spore suspension decanted into a sterile McCartney bottle. Use a loopful of spore suspension to streak across the Cellophane.

Table 3. Germination of spores on Cellophane placed over colonies of the same species (Robinson, unpublished data)
Each figure represents the percentage germination for a population of 100 spores after 24 h on Cellophane over the centre of the colony under test or over sterile S agar (control).

		Colony age (days)				
		1	2	3	4	6
Rhizopus stolonifer	Control	82	80	90	–	–
	Over colony	22[a]	0[a]	0[a]	–	–
Rhizopus arrhizus	Control	32	29	25	–	–
	Over colony	34	0[a]	0[a]	–	–
Cunninghamella elegans	Control	90	89	90	99	–
	Over colony	72[a]	65[a]	4[a]	8[a]	–
Geotrichum candidum	Control	100	100	100	100	98
	Over colony	98	97	90[a]	90[a]	2[a]

[a]Value is significantly different from the corresponding control value at $P < 0.05$.

lysing. After incubation of the drops for 24 h it will be noted that many of the propagules will have germinated and that a healthy mycelium has begun to develop. The staled liquid can also support germination and growth even when considerably diluted. Experiments of this nature have frequently been interpreted as indicating that failure of spores to germinate in a staled culture is not due to insufficient nutrients. It has also been concluded from similar experiments that labile germination- and growth-inhibitory substances are produced by the resident mycelium in a staled culture. However, in the intact staled culture there would be more competition for the available nutrients than in a drop of staled liquid containing a few spores and having a lower hyphal density than the intact culture.

Experiment (2) is an example of an assay system to detect the production of an inhibitory effect by fungal colonies on spore germination. The presence of Cellophane in this experiment enables the test spores to be readily observed by direct microscopic examination of the assay system. The fact that the spores fail to germinate indicates one or more of the following possibilities:

(1) A germination-inhibitory metabolite is produced by the colony and diffuses through the Cellophane to the test spores.

(2) In the case of spores from species which have a nutrient requirement for germination, sufficient nutrients may not be available to the spores on account of competition from the colony beneath. Sporangiospores of *Rhizopus arrhizus* and *Rhizopus stolonifer* require a source of carbon and nitrogen for germination but spores of *Cunninghamella elegans* and *Geotrichum candidum* will germinate in distilled water.

(3) Factors essential for germination of the test spores, and which normally may leak or be leached from the spores, may be utilized by the colony beneath. In this way a deficiency may occur in the spores due to the imposition of a steep diffusion

gradient between spores and colony which results in a failure to germinate in the presence of the colony.

Germination of the test spores over different portions of the colony can also be recorded. It will be observed that the most marked inhibition of germination of the spore population is over the centre of the colony with the effect diminishing as the margin of the colony is approached. The development of the inhibitory effect with colony age is shown in Table 3. Seeded Cellophane strips which have had spore germination inhibited by the colony beneath can be transferred to sterile S agar and the resultant behaviour of the spores noted. The spores will usually germinate which proves that the inhibitory effect is not permanent. Other experiments which could be carried out consist of leaving spores for varying periods (1–7 days) over a colony and testing for germination on transfer to S agar.

8
Mycostasis

Mycostasis (or fungistasis as it is sometimes called) concerns the behaviour of fungal spores in soil. There may be as many as half a million fungal spores in one gram of soil (see Table 4) and the numbers of fungal spores appear to be less affected by the soil pH than do the numbers of bacteria and actinomycetes present. It is important to find out the form in which fungi are present in the soil; are they present as actively growing vegetative hyphae or as spores? There are techniques for answering this question and surprisingly the more common soil species are present most of the time as spores. In so much as many of the spores appear to be ungerminated in apparently favourable conditions they are not behaving normally and the phenomenon is referred to as mycostasis.

The discovery that many spores fail to germinate in soil was made comparatively recently by Dobbs and Hinson (1953). Spores of *Penicillium nigricans* were added to soil and failed to germinate. Simple assay methods for demonstrating mycostasis were developed (Fig. 33) and it was observed that mycostasis was not so marked in soil samples which were taken from well below the soil surface. Also, when soils were autoclaved or mixed with charcoal the inhibitory properties were lost. This leads to the first important conclusion – that the effect is of biological origin, i.e. due in some way to the various living organisms in the soil. Dobbs and Hinson favoured the idea that an inhibitor of spore germination was produced in the soil and that the inhibitor could not be of large molecular weight as it was able to diffuse through Cellophane. The inhibitor was also considered to be unstable in water.

The cause of mycostasis has puzzled scientists ever since. There are two main possibilities: first that the inhibition, as suggested by Dobbs and Hinson, is due to the production of diffusible inhibitory metabolites by soil microorganisms; and second that the inhibition results from a lack (possibly due to competition by other microorganisms) of essential substances required for spore germination.

The first possibility suffered initially from failure to isolate the inhibitors responsible. However, recent research has demonstrated volatile inhibitors of spore germination in soil and more will be said about this later. What about the second possibility? It is feasible that spores may act as nutrient sources in the soil and stimulate the soil microflora to multiply locally and compete for available nutrients. It is known that fungal spores (like seeds of higher plants) 'leak' nutrients and that when spores (dead or alive) are added to soil the soil shows an increase in

Table 4. Number of organisms g^{-1} in a fertile agricultural soil. Reproduced, by permission of Hutchinson Publishing Group Limited, from A. Burges, *Microorganisms in the Soil* (1958)

Bacteria	2.5×10^9	(Direct count)
Bacteria	1.5×10^7	(Dilution plate)
Actinomycetes	7×10^5	
Fungi	4×10^5	
Algae	5×10^4	
Protozoa	3×10^4	

oxygen uptake. Up to ten times the normal number of bacterial colonies may develop from these amended soils. Plastic film can be pressed onto soil surfaces to which spores have been added; a stimulation of the microflora is noted later when the film is removed and the adhering spores examined with the aid of a microscope. It is assumed that the increased numbers of bacteria, etc. compete for soil nutrients and result in a deficiency for the spores. Certainly there is good evidence that nutrient deprivation resulting from the microbial nutrient sink is a factor primarily responsible for mycostasis in certain soil types (Bristow and Lockwood, 1975).

The idea that mycostasis may be in part due to inhibitors has received a fresh impetus with the demonstration by Hora and Baker (1970; 1972) that volatile

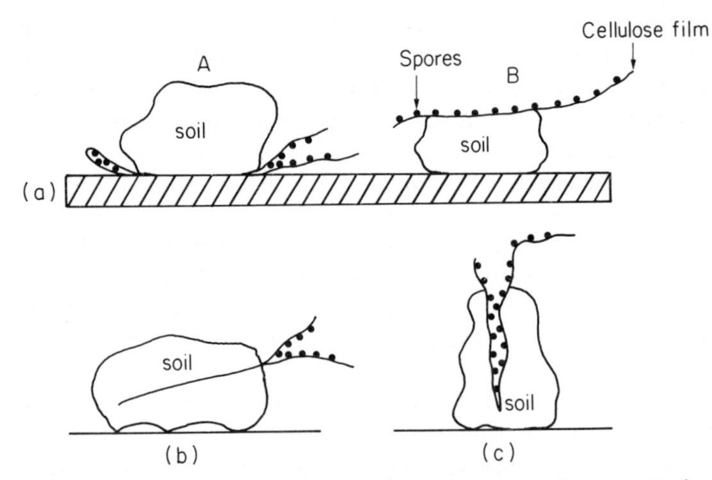

Fig. 33. The cellulose-film test for assaying soil mycostasis. (a) Open film test; a square-shaped portion of film is smeared with spores and folded with the spores inside. The folded film is pressed under a lump of moist soil against a glass slide (A) then the lump of soil is inverted and the film carefully opened up (B). (b) Closed film test; the folded film is pressed between two lumps of soil. (c) Partly-open film test; a closed film is pulled slightly apart to admit air. The free margins of the folds act as controls for the open and partly-open tests. The closed form of test is the most sensitive and imitates most closely the conditions of moisture and aeration which exist in soil. (Dobbs *et al.*, 1960)

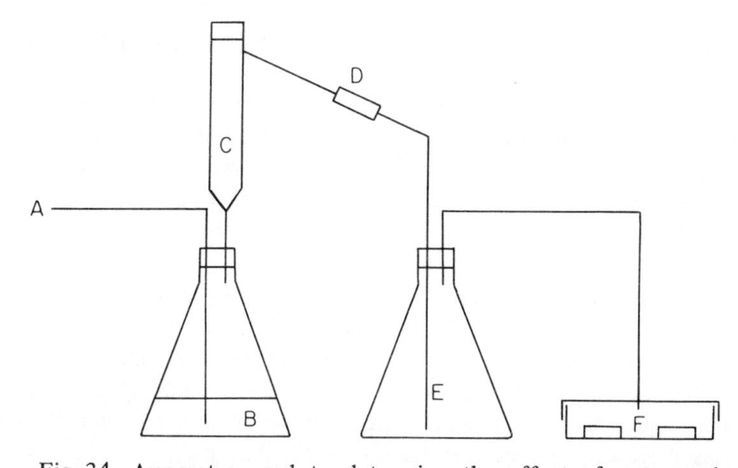

Fig. 34. Apparatus used to determine the effect of water and activated charcoal on the inhibition of conidial germination by a volatile fungistatic factor. Air applied at A at a rate of 5 to 7 ml min^{-1} was washed and humidified in water (B) and passed into a tube (C) containing 20 g soil. In some experiments the arm of the tube (D) was filled with activated charcoal. The air was then passed through a flask (E) (which in some experiments had water added) and finally over spore suspensions seeded on agar discs in a Petri dish (F) (Romine and Baker, 1972). Reproduced by permission of the American Phytopathological Society

inhibitors of spore germination are produced by alkaline soils (pH 7.5 and higher). A simple experimental system for detecting these metabolites is illustrated in Fig. 34 (Romine and Baker, 1972). Such soils lose activity when autoclaved and only produce volatile mycostatic factors when reinoculated with the actinomycetes formerly present. Reinoculation with the species of fungi and/or bacteria previously present does not result in production of the volatile inhibitors. Two volatile mycostatic factors have recently been characterized as ethylene (Smith, 1973) and ammonia (Ko *et al.*, 1974). Although several workers have proposed that ethylene functions as a regulator of soil biological cycles by reducing or inhibiting fungal growth, Archer (1976) has shown that ethylene is not a broad-spectrum fungistatic compound. It is of interest that Balis (1976) reported that ethylene promotes the formation of other volatile, unsaturated derivatives in soil. One of these derivatives, allyl alcohol, was found to be strongly fungistatic. As a result of this work the possibility must be considered that the supposed direct fungistatic effect of soil ethylene is indirectly mediated.

In view of the controversy over the actual cause of mycostasis an attempt has been made (Watson and Ford, 1972) to view the phenomenon as controlled by a complex balance of stimulators and inhibitors of spore germination in soil microenvironments. The stimulators are mostly of biotic origin and may act as nutrients; some may be volatile. The inhibitors are of biotic and abiotic origin. The characterized biotic inhibitors are water soluble and may be volatile in some

Flowsheet 30. Demonstration of mycostasis using *Trichothecium roseum* as test organism

---

Inoculate slopes of malt agar with *T. roseum* and incubate for 1 week at 25 °C. The spores are pink in colour. Add 10 ml sterile distilled water to a slope of the fungus and rub the surface of the culture *gently* with a wire loop to dislodge the spores. Transfer the spore suspension to a sterile McCartney bottle and shake vigorously for 1 min.

Sieve garden soil using a sieve of approximately 3 mm mesh to remove coarse particles. The sieved soil should be moistened with distilled water but care should be taken to ensure that free water is not visible. Fill eight *glass* Petri dishes by packing the sieved soil firmly to a depth of 5 mm and lightly press a clean microscope slide on the soil to leave a slight depression on the surface. Wrap four of the dishes in brown paper and sterilize for 30 min at 15 p.s.i.

↓

Add 1 ml of spore suspension drop-by-drop to each plate so as to cover the slightly depressed area. Incubate plates at 25 °C and after 1, 3, and 7 days (or other suitable intervals) examine a plate from each series (sterile and non-sterile soil) as follows:

↓

Press a *clean* microscope slide on the depressed area of the soil for 20 s. Remove the slide and allow to air-dry. Fix the slide by passing through a bunsen flame three times (this should be done quickly – do not overheat). Stain the slide for 30 s in phenolic rose bengal at 95 °C and then rinse in cold water. Examine as a water mount.

Notes:
Phenolic rose bengal stain: aqueous phenol (5 per cent w/v), 100 ml; rose bengal, 1 g; $CaCl_2$, 0.01 g. The stain should be kept at approximately 95 °C in a glass tube in a boiling water bath.

instances. The abiotic inhibitors include calcium carbonate, iron salts in the soil and, in some cases, the soil pH. Whether or not a spore germinates is considered to be the result of the balance of all these factors in the immediate environment of each spore.

Lockwood (1977) has expressed the opinion that the fungistatic substances-stimulator hypothesis is not consistent with the general uniformity of fungistasis as expressed in nature. He argues that microbial populations in soil must be operating at energy levels only slightly above maintenance requirements. As a result the soil microorganisms, including fungi, must be in an imposed dormancy during most of the time, whether inhibitory factors are present or not. He concludes that inhibitory substances are secondary factors in fungistasis which results primarily from chronic substrate deficiency.

Flowsheet 30 describes an experiment to demonstrate mycostasis. *Trichothecium roseum* has been selected as a test organism because the large, two-celled spores are readily identified when recovered from the soil. Note the failure of the spores to germinate when added to non-autoclaved soil and the germination of spores added to autoclaved soil. Other species of fungi can be tested with this technique.

9
Continuous Culture

It may appear presumptuous to introduce continuous culture as a topic in a book which is claimed not to include experiments which require elaborate apparatus. There are two reasons for doing so. First, there is a growing interest among mycologists in this technique as applied to the filamentous fungi and second, much valuable information on fungal morphogenesis, physiology, and biochemistry can be gained by application of the technique. The apparatus necessary for the continuous culture of microorganisms is called a 'chemostat' and chemostats are often elaborate tools which can cost (at present) in the region of £5000 for the larger models. It is possible nevertheless to set up a small inexpensive chemostat which, although somewhat limited in performance, can give useful information concerning fungal growth and development.

A chemostat has the advantage that many environmental factors can be controlled precisely and in this way there are fewer variables to contend with than in experiments carried out with plate cultures of fungi or with batch liquid cultures. The chemostat consists of a culture of a microorganism to which fresh growth medium is continuously added and from which the culture displaced by this fresh medium is continuously drained away – hence the term 'continuous culture'. In a sophisticated apparatus the temperature, aeration, pH, agitation, and input of fresh medium are controlled to a high degree of accuracy with a resultant increase in cost.

An advanced knowledge of the theory of the chemostat is not essential for the experiments suggested in this section and neither would it be appropriate in this text. It is important, though, to understand some of the fundamental principles and these are outlined only to the extent necessary for setting up and interpreting the experiments in Flowsheet 31. In a chemostat the growth of a microorganism is usually controlled by the concentration of the carbon source (usually glucose) in the growth medium, all other nutrients are in excess. Growth is referred to as 'glucose-limited' under such conditions. The specific growth rate of the microorganism is controlled by the rate at which fresh medium is added to the culture. The faster the medium is added the greater the specific growth rate of the organism. The rate at which the medium is added to a culture is referred to as the dilution rate (D). $D = f/v$, where f = the flow rate (ml h^{-1}) of the medium added and v = the volume (ml) of the culture. Thus for a flow rate of 100 ml h^{-1} and a culture volume of 100 ml, $D = 1$. It is a feature of the chemostat that starting from any initial values of substrate concentration and concentration of organisms (dry weight of organisms/unit volume) the system inevitably adjusts itself to a stable condition known as 'steady state'. Consider, for example, a system which has just been inoculated with a small number of organisms. In this situation the substrate concentration (s) in the culture vessel is nearly equal to the substrate concentration in the inflowing medium and the specific growth rate (μ) of the culture will exceed the dilution rate (D). There will be an increase in the concentration of organisms but, owing to the resultant fall in s, μ will decrease and eventually become equal to D. At this point the rate of substrate addition is just balanced by the combined rates of substrate consumption and loss; the system shows no further tendency to change. Small fluctuations from the steady-state values will set up opposite reactions which will restore equilibrium.

There is considerable disagreement among scientists on the merits of inexpensive chemostats. Many inexpensive chemostats will obviously lack the degree of control possible on more expensive equipment. The degree of control required will of course depend on the nature of the experiments to be carried out. There are several publications which describe the construction of inexpensive chemostats and Primrose (1975) gives details of such a chemostat which is constructed from standard glassware. In this particular chemostat the air supply to the culture vessel (a large test-tube) serves to agitate the culture as well as to aerate it. This raises an important issue. Although such a system works well with microorganisms such as bacteria or non-filamentous yeasts, filamentous fungi require more efficient agitation than would be provided by aeration alone. Ideally, a culture vessel for filamentous fungi should be equipped with an impeller to ensure efficient mixing of the culture.

There are small chemostats available commercially which can be adapted for the culture of filamentous fungi. These are still relatively expensive – £500 to £1000 – but a wide variety of experiments can be carried out with them. *Geotrichum candidum* is a suitable filamentous fungus for continuous-culture work as it tends to fragment spontaneously in liquid culture. However, there are problems with this organism. *G. candidum* has a marked tendency, as do several other fungi, to grow tenaciously on glass and metal surfaces such as would be found inside any culture vessel. The problem is more severe at high dilution rates. There is no complete answer but spraying exposed glass and metal surfaces inside the culture vessel with polytetrafluoroethylene (PTFE) and adding an anti-foam agent such as polypropylene glycol 2000 (Shell Chemicals U.K. Ltd., London, England) to the medium does tend to reduce this effect. Fungi are also notorious for growing along the tube which carries the medium into the culture vessel and along the tube which carries the culture into the harvest or overflow vessel. The first problem can be countered by one or more drip-feed flow breakers in the nutrient feed line. The second problem can be countered by having a communal overflow line for both the displaced culture and the escape of air from the culture. The passage of air through this line tends to minimize build-up of fungus material on the wall of the tube. The problem can also be tackled by the provision of a heated element around the exit point of the glass overflow tube from the culture vessel. There are advantages in

Flowsheet 31. Growth of *Geotrichum candidum* in continuous culture

---

Prepare 1 l of the following culture medium ($g\,l^{-1}$): KH_2PO_4, 3.4; $Na_2HPO_4 \cdot 12H_2O$, 8.9; $(NH_4)_2SO_4$, 6; $MgSO_4 \cdot 7H_2O$, 0.25; $CaCl_2$, 0.05; $ZnSO_4 \cdot 7H_2O$, 0.2; $MnSO_4 \cdot 7H_2O$, 0.02; $CuSO_4 \cdot 5H_2O$, 0.005; $FeSO_4 \cdot 7H_2O$, 0.1; Na_2SO_4, 0.5; $NaMoO_4 \cdot 2H_2O$, 0.05; disodium ethylenediaminetetracetic acid (EDTA), 0.6. Prepare separately, and sterilize, 0.5 l of a glucose solution (9 $g\,l^{-1}$). Add 8 l of distilled water to 0.5 l of the mineral salt solution and sterilize in a glass container which will serve as a medium reservoir for the chemostat. Add all the sterile glucose solution to the 8.5 l sterile mineral salt solution just before commencing the experiment. This gives 9 l of medium containing glucose at a concentration of 0.5 $g\,l^{-1}$.

Sterilize the culture vessel and connect to a sterile overflow reservoir of suitable capacity (10 l or so). The scale of operations described so far would be suitable for a culture vessel with a working volume of approximately 250–500 ml. Connect the medium supply to the culture vessel and pump in the appropriate working volume of medium. Inoculate the medium with a spore suspension of *G. candidum* and allow to grow for 24 h using the culture vessel as a batch culture vessel. Aerate and agitate very gently during this period. Control the temperature at 30 °C.

↓

Adjust the medium flow rate to the desired level and set the pH control module to keep the culture at pH 6.0. It is usual to commence operation at a relatively low dilution rate such as 0.1 h^{-1} until steady-state is attained. Steady-state conditions are normally indicated by the maintenance of a constant dry weight by the culture. In view of the low biomass of the culture under the growth conditions described it is difficult to determine dry weight directly. As an alternative, the optical density (O.D.) of samples from the culture can be checked from time to time. If the O.D. of the culture remains constant over a 48-h period then it can be assumed that steady-state conditions prevail. As a general rule, steady-state conditions would almost certainly be established 72 h after adjusting the medium flow to a particular dilution rate.

Suggestions for experiments

Adjust the apparatus to obtain steady-state operation at $D = 0.05, 0.1, 0.2, 0.3$, and 0.4 h^{-1}. Take samples from each dilution rate and examine as follows.

1. Look for sporulation, particularly at low dilution rates. Two types of spore are produced by *G. candidum* in continuous culture. One type is the familiar cylinder-shaped arthrospore, the other is more rounded and thick-walled. How does sporulation (and type of spore) vary with dilution rate? Use a haemocytometer to estimate numbers of spores.
2. *G. candidum* grows as separate fragments each consisting of a main hypha with or without lateral branches. (Subapical or apical branching does not occur in

Flowsheet 31. (continued)

---

steady-state continuous culture.) Examine fragments from different dilution rates. With increasing dilution rate the fragments become shorter in length but lateral branches increase in number (Table 5).

3. Determine the dimensions of apical and non-apical cells. With increasing dilution rate the average cell length decreases, cell width increases and cell volume increases (Table 6).
4. Examine a range of fragments from a relatively high dilution rate ($0.4\,h^{-1}$). Determine the total length (main hypha + lateral branches) for some of the fragments with the aid of the microscope, eyepiece micrometer, and x 10 objective. Also, record the number of growing (apical) cells on each fragment that is examined. For each fragment determine the hyphal growth unit (total length of main hypha + branches/number of growing tips). Does the hyphal growth unit vary with fragment size? Would you expect it to? Determine the hyphal growth unit for fragments from other dilution rates.

Notes:

1. Without pH control the pH of the medium falls. It is an obvious advantage to have a pH control module and, for the medium described, this will necessitate a supply of 0.5 M-NaOH linked via the module pump to the culture vessel.
2. The organism can be cultured quite satisfactorily at the low concentrations of nutrients suggested. Apart from the obvious economy in terms of cost of nutrients, the low glucose concentration results in a low biomass of *G. candidum* in the culture vessel. This has the advantage that the culture is mixed more efficiently due to the resultant low viscosity. Also, the culture is so dilute that microscopic observation of samples is facilitated and further dilution is often unnecessary prior to examination.
3. Information on the growth kinetics of the organism can be obtained by determination of the residual glucose concentration (s) in the culture at each of the steady states. A portion of culture filtrate can be assayed for s by the glucose oxidase method (the reagents for this assay are available in kit form from Sigma Chemical Company, Kingston-upon-Thames, Surrey, England) and s^{-1} is plotted against μ^{-1}. From this graph the maximum growth rate (μ_{max}) of the organism can be determined. The k_s value (the residual glucose concentration which occurs when the culture is growing at half its maximum specific growth rate) can also be calculated (Fig. 36).

The maintenance value (m) for an organism in chemostat culture is the quantity of a particular energy-providing substrate (usually glucose) which must be supplied to maintain the biomass of the culture. This is a useful parameter as when the glucose supply drops below the maintenance value differentiation occurs in many microorganisms – cutting off the glucose feed completely results in autolysis. If the residual glucose concentration is determined for each of several dilution rates the amount of glucose consumed/unit weight of culture/unit time can be determined for each dilution rate as the glucose concentration in the medium supplied is already known. Plot the specific utilization rate of glucose [g glucose consumed (g dry weight of culture)$^{-1}$ h^{-1}] on the vertical axis against dilution rate on the horizontal axis. The maintenance value is given by the intercept of the resultant line on the vertical axis, i.e. the value of the specific utilization rate of glucose when $D = 0$ (zero growth).

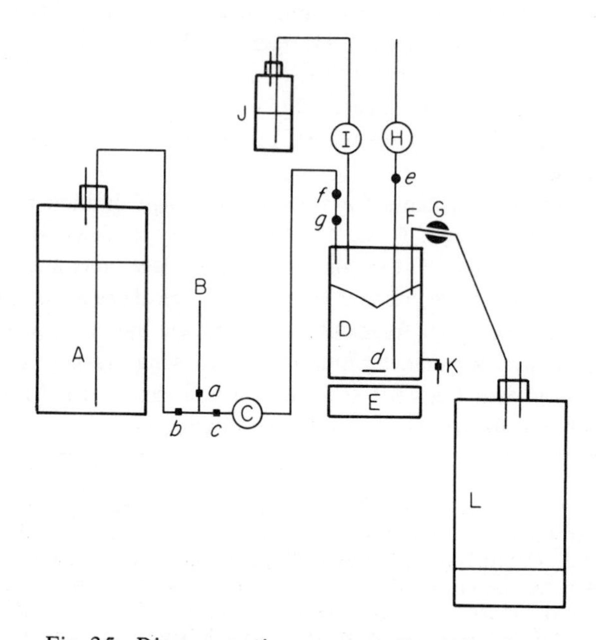

Fig. 35. Diagrammatic representation of a chemostat suitable for the experiments described in Flowsheet 31. A – medium reservoir. B – 10 ml pipette set in medium flow line. During normal operation clip *a* is shut and clips *b* and *c* are open. The pipette can be filled with medium by releasing clip *a*. The medium reservoir can be isolated by closing clip *b* and the medium flow rate is determined by noting the time taken for a particular volume of medium to be pumped from the pipette. C – peristaltic pump in medium flow line. D – culture vessel (pH electrode and heater not shown). E – motor to rotate magnetic stirrer *d*. F – overflow tube (glass) linked to plastic tubing at point beyond heating coil G. H – air pump and filter *e*. I – module to pump alkali from reservoir J in response to a signal that the pH has fallen below the preset value. K – sampling tube. L – harvest vessel. Silicone tubing is used for linking the chemostat components; neoprene tubing is used for the portion of the medium flow line in contact with the rollers of the peristaltic pump. Drip-feed flow breakers are incorporated in the medium flow line at points *f* and *g*

Table 5. Effect of dilution rate on lateral branch formation by *Geotrichum candidum* in glucose-limited chemostat culture at 30 °C (Smith and Robinson, unpublished data)

Dilution rate (h^{-1})	Branch index[a] (%)
0.05	15
0.2	36
0.4	93
0.6	93

[a]The 'branch index' is the percentage of non-apical cells (on the main hypha of a fragment) which produce a lateral branch. Each branch index is based on a count of 300 cells.

having the equipment fitted with a pH control module by which the pH of the medium can be kept at a chosen preset value by the addition of an acid or base.

To summarize, for the chemostat culture of filamentous fungi the culture vessel should be equipped ideally with an impeller and with a pH control module. Some means of temperature control is also essential but to meter accurately the air flow rate is not so important provided air is supplied at a rate sufficient to prevent growth becoming oxygen-limited. Fairly cheap peristaltic pumps can be purchased to meter the flow of medium into the culture vessel and, as an alternative, a Mariotte bottle can be used. Figure 35 illustrates a set-up suitable for the chemostat culture of *G. candidum*.

Table 6. Effect of dilution rate on dimensions of non-apical cells of *Geotrichum candidum* in glucose-limited chemostat culture at 30 °C (Smith and Robinson, unpublished data)

Dilution rate (h^{-1})	Cell length (μm)	Cell width (μm)	Cell volume (μm^3)
0.1	57.35 ± 2.59[a]	2.79 ± 0.09	355.21 ± 31.43
0.25	47.35 ± 5.56	2.82 ± 0.07	312.40 ± 25.01
0.4	44.15 ± 2.69	3.40 ± 0.09	402.73 ± 31.93
0.6[b]	38.90 ± 1.99	4.29 ± 0.14	564.03 ± 41.36

[a]95% confidence intervals.
[b]Steady-state conditions could not be attained at $D = 0.60\ h^{-1}$ and these measurements were taken during 'wash-out'. At 30 °C the maximum specific growth rate (μ_{max}) = $0.50\ h^{-1}$.

In Tables 5 and 6 the results were obtained using a modified modular fermenter with a working volume of 500 ml (A. Gallenkamp & Co. Ltd., Technico House, London, England) using the methods described in Flowsheet 31. Similar results were obtained using a medium with the same constituents but with each ingredient at eighteen times the concentration in Flowsheet 31 and a larger chemostat with a working volume of 3 l (Type LHE 1/1000 fermenter, manufactured by L. H. Engineering Co. Ltd., Stoke Poges, Bucks., England).

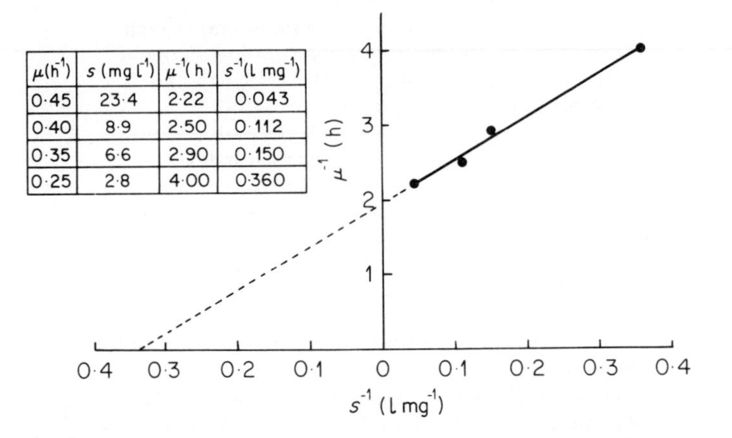

μ (h^{-1})	s (mg l^{-1})	μ^{-1} (h)	s^{-1} (l mg^{-1})
0·45	23·4	2·22	0·043
0·40	8·9	2·50	0·112
0·35	6·6	2·90	0·150
0·25	2·8	4·00	0·360

Fig. 36. Graph of s^{-1} and μ^{-1} for *Geotrichum candidum* growing under glucose-limited conditions in a chemostat at 30 °C. The line of best fit (determined by regression analysis) gives μ_{max} (determined from the intercept on the μ^{-1} axis) as 0.5 h^{-1} and k_s (determined from the intercept on the s^{-1} axis) as 2.94 mg l^{-1}. Each point is the mean of six determinations (duplicate estimations on three separate occasions). (Smith and Robinson, unpublished data)

Flowsheet 31 gives details of some experiments which can be carried out with *G. candidum* using a chemostat. An excellent account of the theory involved in continuous culture is given by Herbert *et al.* (1956).

References

Allen, P. J. (1955). The role of a self-inhibitor in the germination of rust uredospores. *Phytopathology,* **45**, 259–266.

Allen, P. J. (1972). Specificity of the *cis*-isomers of inhibitors of uredospore germination in the rust fungi. *Proceedings of the National Academy of Sciences of the United States of America,* **69**, 3497–3500.

Archer, S. A. (1976). Ethylene and fungal growth. *Transactions of the British Mycological Society,* **67**, 325–326.

Balis, C. (1976). Ethylene-induced volatile inhibitors causing soil fungistasis. *Nature,* **259**, 112–114.

Banbury, G. H. (1954). Processes controlling zygophore formation and zygotropism in *Mucor mucedo* Brefeld. *Nature,* **173**, 499–500.

Barksdale, A. W. (1963). The role of hormone A during sexual conjugation in *Achlya ambisexualis. Mycologia,* **55**, 627–632.

Barksdale, A. W. (1966). Segregation of sex in the progeny of selfed heterozygote of *Achlya bisexualis. Mycologia,* **58**, 802–804.

Barnett, H. L. and Lilly, V. G. (1947). The effects of biotin upon the formation and development of perithecia, asci, and ascospores by *Sordaria fimicola* Ces. & de Not. *American Journal of Botany,* **34**, 196–204.

Barron, G. L. (1968). *The Genera of Hyphomycetes from Soil.* The Williams and Wilkins Company, U.S.A.

Blakeman, J. P. (1969). Self-inhibition of germination of pycnidiospores of *Mycosphaerella ligulicola* in relation to the temperature of their formation. *Journal of General Microbiology,* **57**, 159–167.

Blakeslee, A. F. (1904). Sexual reproduction in the *Mucorineae. Proceedings of the National Academy of Sciences of the United States of America,* **40**, 205–319.

Bristow, P. R. and Lockwood, J. L. (1975). Soil fungistasis: role of the microbial nutrient sink and of fungistatic substances in two soils. *Journal of General Microbiology,* **90**, 147–156.

Brown, W. (1925). Studies on the genus *Fusarium.* II. An analysis of factors which determine the growth forms of certain strains. *Annals of Botany,* **39**, 373–408.

Buller, A. H. R. (1933). *Researches on Fungi*, Vol. 5. Longman, London.

Burges, A. (1958). *Micro-organisms in the Soil.* Hutchinson and Co. Ltd.

Burnett, J. H. (1968). *Fundamentals of Mycology*. Edward Arnold, London.

Butler, G. M. (1961). Growth of hyphal branching systems in *Coprinus disseminatus. Annals of Botany,* **25**, 341–352.

Caglioti, L., Cainelli, G., Camerino, B., Mondelli, R., Prieto, A., Quilico, A., Salvatori, T. and Selva, A. (1967). The structure of trisporic acid-C acid. *Tetrahedron*, Supplement 7, 175–187.

Cainelli, G., Grasselli, P. and Selva, A. (1967). Struttura dell'acido trisporico B. *Chimica e l'industria (Milano),* **49**, 628–629.

Clark, J. F. (1902). On the toxic properties of some copper compounds with special reference to Bordeaux mixture. *Botanical Gazette*, **33**, 26–48.

Cornforth, J. W., Ryback, G., Robinson, P. M. and Park, D. (1971). Isolation and characterization of a fungal vacuolation factor (bikaverin). *Journal of the Chemical Society (C)*, 2786–2788.

Craig, G. D. and Gull, K. (1977). Stipe elongation in *Agaricus bisporus*. *Journal of General Microbiology*, **102**, 337–347.

Dade, H. A. and Gunnell, J. (1969). *Classwork with Fungi*. Commonwealth Mycological Institute, Kew.

Dobbs, C. G. and Hinson, W. H. (1953). A widespread fungistasis in soils. *Nature*, **172**, 197–199.

Dobbs, C. G., Hinson, W. H. and Bywater, J. (1960). Inhibition of fungal growth in soils. In: *The Ecology of Soil Fungi*, pp. 130–147. Liverpool University Press.

Ekundayo, J. A. (1966). Further studies on germination of *Rhizopus arrhizus*. *Journal of General Microbiology*, **42**, 283–291.

Ekundayo, J. A. and Carlile, M. J. (1964). The germination of sporangiospores of *Rhizopus arrhizus*; spore swelling and germ-tube emergence. *Journal of General Microbiology*, **35**, 261–269.

Fischer, F. G. and Werner, G. (1955). Eine Analyse des Chemotropismus einige Pilze, insbesondere der Saprolegniaceen. *Hoppe-Seyler's Zeitschrift für physiologische Chemie*, **300**, 211–236.

Fulton, H. R. (1906). Chemotropism of fungi. *Botanical Gazette*, **41**, 81–108.

Gottlieb, D. (1971). Limited growth in fungi. *Mycologia*, **63**, 619–629.

Graves, A. H. (1916). Chemotropism in *Rhizopus nigricans*. *Botanical Gazette*, **62**, 337–369.

Grove, S. N. and Bracker, C. E. (1970). Protoplasmic organization of hyphal tips among fungi: vesicles and spitzenkörper. *Journal of Bacteriology*, **104**, 989–1009.

Gull, K. (1975). Mycelial branch patterns of *Thamnidium elegans*. *Transactions of the British Mycological Society*, **64**, 321–324.

Gull, K. and Trinci, A. P. J. (1971). Fine structure of spore germination in *Botrytis cinerea*. *Journal of General Microbiology*, **68**, 207–220.

Hawker, L. E. and Abbott, P. McV. (1963). An electron microscope study of maturation and germination of sporangiospores of two species of *Rhizopus*. *Journal of General Microbiology*, **32**, 295–298.

Hawker, L. E., Thomas, B. and Beckett, A. (1970). An electron microscope study of structure and germination of conidia of *Cunninghamella elegans* Lendner. *Journal of General Microbiology*, **60**, 181–189.

Henderson Smith, J. (1924). On the early growth rate of the individual fungus hypha. *New Phytologist*, **23**, 65–78.

Herbert, D., Elsworth, R. and Telling, R. C. (1956). The continuous culture of bacteria; a theoretical and experimental study. *Journal of General Microbiology*, **14**, 601–622.

Hora, T. S. and Baker, R. (1970). Volatile factor in soil fungistasis. *Nature*, **225**, 1071–1072.

Hora, T. S. and Baker, R. (1972). Extraction of a volatile factor from soil inducing fungistasis. *Phytopathology*, **62**, 1475–1476.

Horr, W. H. (1936). Utilization of galactose by *Aspergillus niger* and *Penicillium glaucum*. *Plant Physiology*, **11**, 81–99.

Hughes, S. J. (1953). Conidiophores, conidia, and classification. *Canadian Journal of Botany*, **31**, 577–659.

Jaffe, L. F. (1966). On autotropism in *Botrytis*: measurement technique and control by CO_2. *Plant Physiology*, **41**, 303–306.

Katz, D., Godstein, D. and Rosenberger, R. F. (1972). Model for branch initiation in *Aspergillus nidulans* based on measurements of growth parameters. *Journal of Bacteriology,* **109**, 1097–1100.

Khan, S. R. (1975). Wall structure and germination of spores in *Cunninghamella echinulata. Journal of General Microbiology,* **90**, 115–124.

Ko, W. H., Lockwood, J. L., Hora, F. K. and Herlicksa, F. (1974). Isolation and identification of a volatile fungistatic substance from alkaline soil. *Phytopathology,* **64**, 1398–1400.

Leopold, L. B. (1971). Trees and streams: the efficiency of branching patterns. *Journal of Theoretical Botany,* **31**, 339–354.

Lingappa, B. T. and Lingappa, Y. (1965). Effects of nutrients on self-inhibition of germination of conidia of *Glomerella cingulata. Journal of General Microbiology,* **41**, 67–75.

Lingappa, B. T. and Lingappa, Y. (1966). The nature of self-inhibition of germination of conidia of *Glomerella cingulata. Journal of General Microbiology,* **43**, 91–100.

Lockwood, J. L. (1977). Fungistasis in soils. *Biological Reviews,* **52**, 1–43.

Macko, V., Staples, R. C., Allen, P. J. and Renwick, J. A. A. (1971). Identification of the germination self-inhibitor from wheat stem uredospores. *Science,* **173**, 835–836.

Mesland, D. A. M., Huisman, J. G. and van den Ende, H. (1974). Volatile sex hormones in *Mucor mucedo. Journal of General Microbiology,* **80**, 111–117.

Müller, D. and Jaffe, L. F. (1965). A quantitative study of cellular rheotropism. *Biophysical Journal,* **5**, 317–335.

Musgrave, A., Ero, L., Scheffer, R. and Oehlers, E. (1977). Chemotropism of *Achlya bisexualis* germ hyphae to casein hydrolysate and amino acids. *Journal of General Microbiology,* **101**, 65–70.

Park, D. (1961). Morphogenesis, fungistasis, and cultural staling in *Fusarium oxysporum* Snyder and Hansen. *Transactions of the British Mycological Society,* **44**, 377–390.

Park, D. and Robinson, P. M. (1964). Isolation and bioassay of a fungal morphogen. *Nature,* **203**, 988–989.

Park, D. and Robinson, P. M. (1966). Internal pressure of hyphal tips of fungi, and its significance in morphogenesis. *Annals of Botany,* **30**, 425–439.

Park, D. and Robinson, P. M. (1967). A fungal hormone controlling internal water distribution normally associated with cell aging in fungi. *Symposia of the Society for Experimental Botany,* **21**, 323–336.

Parker, R. E. (1973). *Introductory Statistics for Biology*. Edward Arnold Ltd.

Plempel, M. (1963). Die chemischen Grundlagen der Sexualreaktion bei Zygomyceten. *Planta,* **59**, 492–508.

Plomley, N. J. B. (1959). Formation of the colony in the fungus *Chaetomium. Australian Journal of Biological Sciences,* **12**, 53–64.

Primrose, S. B. (1975). Inexpensive chemostat constructed from standard glassware. *Laboratory Practice,* 88–89.

Raper, J. R. (1940). Sexual hormones in *Achlya*. II. Distance reactions, conclusive evidence for a hormonal coordinating mechanism. *American Journal of Botany,* **27**, 162–173.

Raper, J. R. (1952). Chemical regulation of sexual processes in the thallophytes. *Botanical Review,* **18**, 447–545.

Righelato, R. C., Trinci, A. P. J., Pirt, S. J. and Peat, A. (1968). The influence of maintenance energy and growth rate on the metabolic activity, morphology, and conidiation of *Penicillium chrysogenum. Journal of General Microbiology,* **50**, 399–412.

Robertson, N. F. (1965). The mechanism of cellular extension and branching. In: *The Fungi*, Vol. I (G. C. Ainsworth and A. S. Sussman, eds.), pp. 615–623. Academic Press, New York.

Robinson, P. M. (1973a). Autotropism in fungal spores and hyphae. *Botanical Review,* **39**, 367–384.

Robinson, P. M. (1973b). Oxygen – positive chemotropic factor for fungi? *New Phytologist,* **72**, 1349–1356.

Robinson, P. M. and Griffith, P. J. (1977). Effect of restricted aeration on chemotropism, morphogenesis and polarity of lateral branch induction in *Geotrichum candidum* Link ex Pers. *Transactions of the British Mycological Society,* **68**, 311–314.

Robinson, P. M. and Park, D. (1965). The production and quantitative estimation of a fungal morphogen. *Transactions of the British Mycological Society,* **48**, 561–571.

Robinson, P. M., Park, D. and Graham, T. A. (1968). Autotropism in fungal spores. *Journal of Experimental Botany,* **19**, 125–134.

Romine, M. and Baker, R. (1972). Properties of a volatile fungistatic factor in soil. *Phytopathology,* **62**, 602–605.

Smith, A. M. (1973). Ethylene as a cause of soil fungistasis. *Nature,* **246**, 311–313.

Snedecor, G. W. and Cochran, W. G. (1967). *Statistical Methods.* Iowa State University Press, U.S.A.

Stadler, D. R. (1952). Chemotropism in *Rhizopus nigricans*: the staling reaction. *Journal of Cellular and Comparative Physiology,* **39**, 449–474.

Trinci, A. P. J. (1969). A kinetic study of the growth of *Aspergillus nidulans* and other fungi. *Journal of General Microbiology,* **57**, 11–24

Trinci, A. P. J. (1970). Kinetics of apical and lateral branching in *Aspergillus nidulans* and *Geotrichum lactis. Transactions of the British Mycological Society,* **55**, 17–28.

Trinci, A. P. J. (1972). Culture turbidity as a measure of mould growth. *Transactions of the British Mycological Society,* **58**, 467–473.

Trinci, A. P. J. (1974). A study of the kinetics of hyphal extension and branch initiation of fungal mycelia. *Journal of General Microbiology,* **81**, 225–236.

Trinci, A. P. J. and Whittaker, C. (1968). Self-inhibition of spore germination in *Aspergillus nidulans. Transactions of the British Mycological Society,* **51**, 594–596.

van den Ende, H. (1968). Relationship between sexuality and carotene synthesis in *Blakeslee trispora. Journal of Bacteriology,* **96**, 1298–1303.

van den Ende, H. (1976). *Sexual Interactions in Plants.* Academic Press.

Watson, A. G. and Ford, E. J. (1972). Soil fungistasis – a reappraisal. *Annual Review of Phytopathology,* **10**, 327–348.

Went, F. W. (1936). Algemeine Betrachtungen über das Auxin-Problem. *Biologische Zentralblatt,* **56**, 449–463.

Wortmann, J. (1887). Zur Kenntniss der Reizbewegungen. *Botanische Zeitung,* **45**, 801–812.

Appendix

ABBREVIATIONS

cm	centimetre(s)
mm	millimetre(s)
μm	micrometre(s)
g	gram(s)
mg	milligram(s)
μg	microgram(s)
l	litre(s)
ml	millilitre(s)
μl	microlitre(s)
h	hour(s)
min	minute(s)
s	second(s)
°C	degree(s) Celsius
Σ	sum of
χ^2	chi^2
+	plus (add)
×	times (multiply)
v/v	volume/volume
w/v	weight/volume
M	molar
mM	millimolar
b.p.	boiling point
p.s.i.	pounds per square inch
per cent	percentage
%	percentage (in tabular setting)
$<$	less than

FUNGI USED IN EXPERIMENTS

Achlya ambisexualis J. R. Raper
Agaricus bisporus (Lange) Imbach
Alternaria tenuis Nees
Aspergillus giganteus Wehmer
Aspergillus janus Raper & Thom
Aspergillus niger van Tieghem
Aspergillus repens (Corda) Saccardo
Aspergillus rugulosus Thom & Raper
Botrytis cinerea Persoon ex Persoon
Candida albicans (Robin) Berkout
Chaetomium globosum Kunze ex Fries
Cladosporium herbarum (Persoon) Link ex S. F. Gray
Cunninghamella elegans Lendner
Eremascus albus Eidam
Fusarium oxysporum Schlechtendahl
Geotrichum candidum Link ex Persoon
Lycoperdon perlatum Persoon
Mucor mucedo Brefeld
Mucor plumbeus Bonorden
Neocosmospora vasinfecta E. F. Smith
Neurospora crassa Shear & Dodge
Penicillium chrysogenum Thom
Penicillium digitatum Saccardo
Penicillium frequentans Westling
Penicillium purpurogenum Stoll
Rhizopus arrhizus Fischer
Rhizopus stolonifer (Ehrenberg ex Fries) Lind
Rhodotorula rubra (Demme) Lodder
Saccharomyces cerevisiae Hansen
Saprolegnia ferax (Gruithuisen) Thuret
Schizosaccharomyces octosporus Beijerinck
Scopulariopsis brevicaulis (Saccardo) Bainier
Serpula lacrymans (Wulfen ex Fries) Schroeter
Sordaria fimicola (Roberge) Cesati & de Notaris
Sporobolomyces odorus Derx
Tilletiopsis minor Nyland
Trichoderma viride Persoon ex S. F. Gray
Trichosporon cutaneum (de Beurmann, Gougerot & Vaucher) Ota
Trichothecium roseum (Persoon) Link ex S. F. Gray

BACTERIA USED IN EXPERIMENTS

Bacillus cereus Frankland & Frankland
Bacillus subtilis Cohn emend. Prazmowski
Staphylococcus aureus Rosenbach

LABORATORY TECHNIQUE

General notes

1. Before starting any of the experiments or procedures it is important that all the steps involved are understood.
2. It is preferable to wear a clean laboratory coat. Failing this a clean cotton garment (e.g. shirt, blouse) is better than a jacket, jersey or jumper which can carry a high level of microbial contamination.
3. Develop the habit of handling all sterilized equipment so as to keep it sterile, e.g. if an autoclaved pipette is to be used, remove it from the container by the plugged, mouthpiece end, and handle this end only.
4. Do not leave lids or closures off culture vessels longer than necessary. Organisms may get in or out – either way can lead to trouble.
5. Do not leave spilled cultures or culture media on the bench or floor. Clean them up properly. Neglecting this can lead to a source of contaminated cultures.
6. Equipment which has contained living organisms should be sterilized as soon as possible, even though only a few of the species featured in this book are potentially pathogenic. Pipettes and slides should be put in a cylinder of Lysol or Roccal on the bench. Unwanted glass Petri dishes and other culture vessels should be put in an autoclave bucket ready for sterilizing.
7. Always label culture vessels clearly.

Use of loops and needles for inoculating

1. Heat loops and needles to red heat over a bunsen burner flame and allow to cool (for about 10 s) immediately before using.
2. Always flame the mouth of bottles, tubes, flasks, etc. when removing metal caps, stoppers or cotton plugs, *and again* before replacing.
3. Hold tubes, etc. in such a way that they are opened for the minimum period of time. Do not place the plug, stopper, or lid on the bench.
4. Learn to work near a lighted bunsen where air-borne particles will be carried upwards in a convection current and will be less likely to settle on the surface of the culture vessel.

CULTURE COLLECTIONS

Most of the organisms listed in this book can be obtained from centres which specialize in the maintenance of culture collections. Although there are numerous small collections distributed in laboratories throughout the world there are only a few collections which are permanent and contain a comprehensive list of species.

Some of the centres from which cultures can be purchased are listed below:

The Commonwealth Mycological Institute (CMI),
The Ministry of Technology,
Ferry Lane,

Kew,
Surrey,
England.

British National Collection of Yeast Cultures
Brewing Industry Research Foundation,
Nutfield,
Surrey,
England.

National Collection of Industrial Bacteria,
Ministry of Technology,
P.O. Box 31,
135 Abbey Road,
Aberdeen,
Scotland.

American Type Culture Collection (ATCC),
12301 Parklawn Drive,
Rockville,
Maryland 20852,
U.S.A.

Centraalbureau voor Schimmelcultures (CBS),
Baarn,
The Netherlands.

Catalogues can be purchased from these institutions and the appropriate fungi or bacteria ordered . If in any doubt as to which strain of a particular species to order it is useful to mention (in your order) what the culture is required for. In this way you are more likely to receive a strain appropriate for your purpose.

There are obvious advantages in maintaining a small culture collection of your own. At Belfast most of our fungal cultures are maintained on cornmeal agar slopes in glass universal bottles (28 ml capacity); yeasts are maintained on malt agar slopes. On cornmeal agar fungi grow slowly and it is only necessary to sub-culture at intervals of six months or so. The screw-capped bottles do have certain disadvantages – they are not proof against mite infestation – but are easy to manipulate. The cultures are stored in metal biscuit tins and, as a deterrent for mites, a tissue impregnated with Kelthane (The Murphy Chemical Co. Ltd., Wheathampstead, St. Albans, Herts., England [*Caution:* observe safety precautions written on container]) is added to each tin. This system works well in practice.

It is always advisable to keep master cultures of any fungi that are obtained. Each fungus can be grown on a slope of cornmeal agar in the usual way and sterile mineral (paraffin) oil added to the culture until the McCartney bottle is almost full. Cultures will survive for many years in this condition at laboratory temperature.

Whenever it is necessary to sub-culture the master culture a little material can be picked up from beneath the oil with a sterile loop and transferred to a slope or plate of solid medium.

The species of bacteria mentioned in this book can be maintained as follows. Fill small glass specimen tubes (approximately 45 mm long x 12 mm diameter, screw-capped) to a depth of 15 mm with nutrient agar at one-half the concentration recommended for normal culture work. Any commercial nutrient agar can be used. Autoclave these small deeps at 15 p.s.i. for 15 min. When the medium has set inoculate by stabbing a long needle through the medium. Incubate at 30 °C for 2 days and, after securing the caps firmly, dip the cap and upper part of the tube in molten paraffin wax to make an air-tight seal. The cultures can be stored at laboratory temperature and need only be sub-cultured at intervals of one year.

TREATMENT OF RESULTS

In any experiment it is important to assess the degree of reliability that can be placed on the results. For this purpose it is convenient to analyse the results by the use of statistics. In addition, there are obvious advantages in designing experiments so that the results will be amenable to a simple statistical analysis. Two typical experimental situations are now discussed with these points in mind.

Analysis of results of experiments in which spore germination is recorded

In many of the experiments described in Chapter 1 an estimate of the germination of a population of spores is required. The data obtained are binomially distributed as the spores which are counted are placed into one of two categories which are mutually exclusive – either the spore germinates or it doesn't germinate.

Suppose spores are seeded on the surface of a medium and each spore in a sample of 100 spores is recorded as either germinated or non-germinated. If after, say, 1.5 h it is found that 70 spores in the sample have germinated this gives an estimate of 70 per cent for germination. How reliable is this estimate? If it is assumed that the spores behave independently of each other, i.e. they do not influence each other's germination, then confidence intervals (per cent) for this estimate can be read directly from a table of confidence intervals for a binomial distribution such as Table 1.4.1 of *Statistical Methods* by Snedecor and Cochran (1967). The confidence interval (95 per cent) is 60–80. It is a very simple procedure to determine the confidence interval for each germination percentage that is recorded and this is to be encouraged. Confidence intervals for samples ranging in size from 10 to 1000 can be determined from the same table.

The following example raises an additional point. Suppose two treatments result in 50 per cent and 69 per cent germination for samples of 100 spores. The respective 95 per cent confidence intervals (per cent) are 40 to 60 and 59 to 78. At first sight a difference of 19 per cent between treatments may appear large but statistically the results do not differ. However, if 50 per cent and 69 per cent germination were recorded for samples of 250 spores then the respective 95 per cent confidence intervals (per cent) would be 44 to 56 and 63 to 75 – the results would be significantly different. The value of large samples has already been mentioned in Chapter 1.

Analysis of results of experiments in which tropic responses of germinating spores are studied

In Section 1.5 tropic responses of germinating arthrospores were studied. If attention is confined to the responses of single arthrospores there are two categories of tropic response, either the germ-tube emerges from the end of the arthrospore towards the stimulus or from the other end. Such experiments are suitable for a χ^2 (chi^2) analysis. An example follows.

Suppose arthrospores are seeded at a density of 5×10^5 arthrospores ml^{-1} in 1 per cent malt agar and a coverslip is placed over a portion of the seeded medium. If the tropic responses of a sample of 100 arthrospores (beneath the coverslip) are recorded it may be found that each of 80 arthrospores germinates from the end nearer the coverslip margin and 20 germinate from the end away from the margin. Do these results reflect a tropic response in the direction of the coverslip margin? The following formula can be used to find out:

$$\chi^2 = \Sigma \frac{(O - E)^2}{E}$$

where O is the observed frequency for a particular response and E is the corresponding expected frequency. In the example above it might be expected that 50 arthrospores (half the observed population) would germinate towards the margin if the coverslip had no effect on the position of germ-tube emergence, i.e. $E = 50$. The analysis is as follows:

Class	Observed	Expected	Deviation
Germinate towards margin	80	50	+30
Germinate away from margin	20	50	–30
(Total)	(100)	(100)	(0)

Using the general formula for χ^2:

$$\chi^2 = \frac{+30^2}{50} + \frac{-30^2}{50} = 36 \text{ with one degree of freedom.}$$

When there is only one degree of freedom, i.e. only two classes (as above), it is necessary to modify the formula for χ^2 by the application of Yates's correction. The modified formula is as follows:

$$\chi^2 = \Sigma(|O - E| - 1/2)^2/E$$

where $|O - E|$ is the positive difference between O and E. When Yates's correction is applied to the example above,

$$\chi^2 = \frac{(30 - 1/2)^2}{50} + \frac{(30 - 1/2)^2}{50} = 34.81 \text{ with one degree of freedom.}$$

From a table of χ^2 it can be determined that the probability of obtaining a value of 34.81 is <0.001 which means that the supposed tropic response is highly significant. The application of Yates's correction does not affect the conclusion in this example but the correction can be important when deviations border on significance.

In this example it was assumed, for convenience, that E would be 50 if the coverslip had no effect on germ-tube orientation. While this may be a reasonable assumption it is advisable, where possible, to determine such values experimentally.

These two examples of statistical analysis represent an elementary attempt to encourage the use of statistics for some of the experiments described in this book. For a more comprehensive treatment of the subject the reader is referred to *Introductory Statistics for Biology* (Parker, 1973) and *Statistical Methods* (Snedecor and Cochran, 1967).

Index

Italicized page numbers refer to figures